Wireless Optical Communications

Wireless Optical Communications

Olivier Bouchet

Series Editor
Pierre-Noël Favennec

First published 2012 in Great Britain and the United States by ISTE Ltd and John Wiley & Sons, Inc.

ISTE Ltd
27-37 St George's Road
London SW19 4EU
UK

www.iste.co.uk

John Wiley & Sons, Inc.
111 River Street
Hoboken, NJ 07030
USA

www.wiley.com

© ISTE Ltd 2012

Library of Congress Cataloging-in-Publication Data

Bouchet, Olivier.
Wireless optical telecommunications / Olivier Bouchet.
p. cm.
Includes bibliographical references and index.
 ISBN 978-1-84821-316-6
 1. Wireless communication systems. 2. Optical communications. I. Title.
 TK5103.2.B69 2012
 621.382'7--dc23
                                                                              2012005891

British Library Cataloguing-in-Publication Data
A CIP record for this book is available from the British Library
ISBN: 978-1-84821-316-6

Printed and bound in Great Britain by CPI Group (UK) Ltd., Croydon, Surrey CR0 4YY

Table of Contents

Foreword. xi
Pierre-Noël FAVENNEC

Acronyms xiii

Introduction. xix

Chapter 1. Light 1

Chapter 2. History of Optical Telecommunications 7

2.1. Some definitions. 7
   2.1.1. Communicate 7
   2.1.2. Telecommunication 7
   2.1.3. Optical telecommunication 8
   2.1.4. Radio frequency or Hertzian waves 8
2.2. The prehistory of telecommunications 8
2.3. The optical aerial telegraph 11
2.4. The code 14
2.5. The optical telegraph 18
   2.5.1. The heliograph or solar telegraph 18
   2.5.2. The night and day optical telegraph 19
2.6. Alexander Graham Bell's photophone 20

**Chapter 3. The Contemporary and the Everyday Life of Wireless
Optical Communication** 25

3.1. Basic principles 25
   3.1.1. Operating principle. 26
      3.1.1.1. Block diagram 26
   3.1.2. The optical propagation 27

3.1.2.1. Line of sight propagation – LOS   27
3.1.2.2. Wide line of sight – WLOS   29
3.1.2.3. Diffusion propagation (DIF) and controlled diffusion   30
3.1.3. Elements of electromagnetics.   31
3.1.3.1. Maxwell's equations in an unspecified medium   32
3.1.3.2. Propagation of electromagnetic waves in an
isotropic medium   34
3.1.3.3. Energy associated to a wave   36
3.1.3.4. Propagation of a wave in a non-homogeneous medium   38
3.1.3.5. Coherent and incoherent waves   38
3.1.3.6. Relations between electromagnetism and
geometrical optics   40
3.1.3.7. The electromagnetic spectrum   43
3.1.3.8. Units and scales   43
3.1.3.9. Examples of sources in the visible and near visible light   47
3.1.3.10. Conclusion.   49
3.1.4. Models for data exchange   50
3.1.4.1. The OSI model   50
3.1.4.2. The DoD model   52
3.2. Wireless optical communication   53
3.2.1. Outdoor wireless optical communication   53
3.2.1.1. Earth-satellite wireless optical communication   53
3.2.1.2. Intersatellite wireless optical communication   54
3.2.1.3. Free-space optic   55
3.2.2. Indoor wireless optical communication   55
3.2.2.1. The remote controller   56
3.2.2.2. The visible light communication   57
3.2.2.3. The IrDA solutions   57
3.2.2.4. The indoor wireless optical network (WON)   57
3.2.3. The institutional and technical ecosystem   59

Chapter 4. Propagation Model   63

4.1. Introduction.   63
4.2. Baseband equivalent model   63
4.2.1. Radio propagation model   64
4.2.2. Model of free-space optical propagation   66
4.2.3. The signal-to-noise ratio.   71
4.3. Diffuse propagation link budget in a confined environment   73
4.3.1. Intersymbol interference.   73
4.3.2. Reflection models.   76
4.3.2.1. Specular reflection.   76
4.3.2.2. Diffuse reflection   76
4.3.2.3. Lambert's model.   77

4.3.2.4. Phong's model   79
4.3.3. Modeling   81

Chapter 5. Propagation in the Atmosphere   85

5.1. Introduction.   85
5.2. The atmosphere   86
5.2.1. The atmospheric gaseous composition   86
5.2.2. Aerosols   87
5.3. The propagation of light in the atmosphere   87
5.3.1. Molecular absorption.   89
5.3.2. Molecular scattering   89
5.3.3. Aerosol absorption   90
5.3.4. Aerosol scattering.   91
5.4. Models.   93
5.4.1. Kruse and Kim models.   93
5.4.2. Bataille's model.   94
5.4.2.1. Molecular extinction   94
5.4.2.2. Aerosol extinction   95
5.4.3. Al Naboulsi's model   95
5.4.4. Rain attenuation.   96
5.4.5. Snow attenuation   97
5.4.6. Scintillation   98
5.5. Experimental set-up   103
5.6. Experimental results.   104
5.6.1. Comparaison with Kruse and Kim model (850 nm)   105
5.6.2. Comparaison with Al Naboulsi's model   105
5.7. Fog, haze, and mist   107
5.8. The runway visual range (RVR)   108
5.8.1. The visibility.   108
5.8.2. Measuring instruments.   110
5.8.2.1. The transmissometer   110
5.8.2.2. The scatterometer   112
5.9. Calculating process of an FSO link availability   114
5.10. Conclusion   116

Chapter 6. Indoor Optic Link Budget.   119

6.1. Emission and reception parameters.   119
6.1.1. Transmission device: parameters.   121
6.1.2. Reception device   125
6.2. Link budget for line of sight communication   128
6.2.1. Geometrical attenuation   128
6.2.2. Optical margin   130

6.2.3. Coverage 130
6.2.4. Reciprocity and not reciprocity of the channel. 131
6.3. Link budget for communication with retroreflectors. 132
6.3.1. Principle of operation 132
6.3.2. Optical budget. 133
6.4. Examples of optical budget and signal-to-noise ratio (SNR). 135
6.4.1. Examples of optical budget 136
6.4.2. Examples of SNR and BER. 139

Chapter 7. Immunity, Safety, Energy and Legislation 141

7.1. Immunity 141
7.1.1. International references 141
7.1.2. Type of laser classes 143
7.1.3. Method for calculation. 146
7.2. The confidentiality of communication 149
7.2.1. Physical confidentiality 149
7.2.2. Numerical solution 150
7.2.2.1. Cryptography. 150
7.2.2.2. Public and secret key cryptography. 151
7.2.2.3. Quantum cryptography 151
7.2.2.4. Quantum telecommunications in free space. 152
7.2.2.5. Non-encrypted connections in confined space 153
7.3. Energy. 153
7.4. Legislation 154
7.4.1. Organization of regulation activities. 154
7.4.2. Regulation of wireless optical equipment. 155

Chapter 8. Optics and Optronics. 157

8.1. Overview 157
8.2. Optronics: transmitters and receivers. 157
8.2.1. Overviews on materials and structures 157
8.2.2. Light sources 160
8.2.2.1. Light-emitting diodes (LEDs) and spontaneous emission. . . . 161
8.2.2.2. White LEDs or visible light communication (VLC) LED. . . . 162
8.2.2.3. The semiconductor laser structure 163
8.2.2.4. Synthesis 165
8.2.3. Optronics receivers. 166
8.2.3.1. Photovoltaic cells 167
8.2.3.2. PIN photodiode. 168
8.2.3.3. Avalanche photodiode 169
8.2.3.4. Metal–semiconductor–metal (MSM) structure 170
8.3. Optics 170

8.3.1. Transmitter optical device.   170
8.3.2. Receiver optical device   171
8.3.3. Optical filtering   174
    8.3.3.1. Spatial filter or diaphragm..   174
    8.3.3.2. Wavelength filters or attenuators   174
8.3.4. Summary.   176

Chapter 9. Data Processing   177

9.1. Introduction.   177
9.2. Modulation   178
9.2.1. On-off keying (OOK) modulation   178
9.2.2. The pulse position modulation   180
9.2.3. The orthogonal frequency-division multiplexing (OFDM)   181
9.2.4. The diversity: MIMO   182
9.2.5. Summary.   184
9.3. The coding   184
9.3.1. Principle and definitions.   184
    9.3.1.1. Principle.   184
    9.3.1.2. Definitions   185
9.3.2. Example of coding   186
    9.3.2.1. Basic codes   186
    9.3.2.2. Block codes.   187
    9.3.2.3. Convolutional codes.   191
9.3.3. Summary.   194

Chapter 10. Data Transmission.   197

10.1. Introduction.   197
10.1.1. Definition.   197
10.1.2. The access methods.   198
    10.1.2.1. Time division multiple access   198
    10.1.2.2. Frequency division multiple access   199
    10.1.2.3. Code division multiple access   199
    10.1.2.4. Carrier sense multiple access.   199
    10.1.2.5. Wavelength division multiple access   199
    10.1.2.6. Space division multiple access.   200
10.1.3. Quality of service parameters   200
10.2. Point-to-point link   201
10.2.1. The remote control   201
10.2.2. Infrared Data Association   203
10.2.3. Visible light communication consortium   206
10.3. Point-to-multipoint data link.   206
10.3.1. IEEE 802.11 IR   206

10.3.2. ICSA – STB50 (IEEE 802.3 – Ethernet)   208
10.3.3. IEEE 802.15.3   209
10.3.4. IEEE 802.15.7   209
10.3.5. Optical wireless media access control   210
10.4. Summary   212

Chapter 11. Installation and System Engineering   213

11.1. Free-space optic system engineering and installation   213
11.1.1. Principle of operation.   213
11.1.2. Characteristics   214
11.1.2.1. Principal parameters   215
11.1.2.2. Secondary parameters   216
11.1.2.3. Examples of FSO systems   216
11.1.3. Implementation recommendations   217
11.1.4. Optic link budget   218
11.1.4.1. Geometrical attenuation concept.   219
11.1.4.2. Link margin concept   219
11.1.5. FSO link availability   220
11.1.5.1. Characteristics   220
11.1.5.2. Results   223
11.1.6. Summary   225
11.2. Wireless optical system installation engineering in limited space . . .   225
11.2.1. Habitat structure.   225
11.2.2. Statistical analysis and coverage area   226
11.2.3. Optical link budget   230
11.2.4. Optimization of indoor wireless optical system   234

Chapter 12. Conclusion.   237

APPENDICES   241

Appendix 1. Geometrical Optics, Photometry and Energy Elements   243

Appendix 2. The Decibel Unit (dB)   257

Bibliography   261

List of Figures   273

List of Tables   277

List of Equations   279

Index   283

Foreword

Modern telecommunication, at least in the vicinity of terminals (TV receivers, computers, recorders, smartphones, network games consoles, e-books, etc.), will be "wireless" and high speed: the physical link will not be a copper wire, or made from fiber, silica, or other, but an electromagnetic wave propagating in free space between one transmitter–receiver and another transmitter–receiver.

The most common physical wireless link is the use of radio, an electromagnetic wave in the range of radio spectrum. It is a well-developed technology, but we can see the limitations in terms of speed (bits per second), frequency, power, electromagnetic compatibility, and electromagnetic pollution among others. Regarding transmission of information, we know that the higher the frequency of the electromagnetic transmitted wave, the higher the speed. Hence, current laboratory studies are looking at communication systems operating at frequencies of gigahertz (GHz) to terahertz (THz) and above. For frequencies beyond terahertz, and particularly in the ranges corresponding to optical waves, infrared or visible light (100–1,000 THz), a communication speed in the range of terabits per second can be achieved.

Because of the laser (invented in 1960) and silica fiber (the potential of silica fiber for telecom applications was demonstrated in 1961), optical telecommunications together with the fantastic progress made in the manufacturing technology of lasers and optoelectronic systems, in parallel to those of silica fibers, have enabled the irreversible development of optical fiber telecommunication. These optical communications have generated intercontinental telecommunications and broadband internet. From basic-oriented research, they have an obvious important societal impact.

Wireless optical communications use the atmosphere as a transmission medium. The ambient atmosphere is much more complex than the fibrous silica in terms of composition, uniformity, and reproducibility. But taking advantage of advanced technologies useful for fiber telecommunications, it gives excellent results for broadband transmitted over short distances and even allows us a glimpse of wireless optical communication with terabits per second, even though today (in 2011) we are using gigabit to the terminal (GTTT) in a limited confined environment.

The atmospheric optical links are always subject to environmental variations (dust, fog, rain, etc.), which can cause temporary performance degradation of the telecommunications system. The propagation properties of optical beams in this environment must provide a good quality of service as in the model of Al Naboulsi *et al.* [NAB 04], based on visibility, the setting that characterizes the opacity of the atmosphere. Using components (LED, laser, photodetectors, etc.) at wavelengths that are non-ionizing photons whose technologies are now mature, in free-space communication over short distances, especially indoors (in rooms), has great potential. The book *Wireless Optical Communications* follows a previous book *Free-Space Optics – Propagation and Communication* [BOU 06] that presented the physics and foundations useful for communications in free space and in limited spaces. Since the last book, great progress has been made on all issues related to a real telecommunications system incorporating channel properties, propagation models, link budgets, and the data processing including coding, modulation, standards, and safety.

This book is designed as an excellent tool for any engineer wanting to learn about wireless optical communications or who is involved in the implementation of real complete systems. Students will find lots of information and useful concepts such as those relating to propagation, optics, and photometry, as well as the necessary information on safety.

This book is written with as an overview of a useful technology for telecommunications. The ideas developed allow us a glimpse of the applications in the field of communication devices by photons. Since the early work of Gfeller in 1979 on optical wireless limited space [GFE 79] or the work of Kintzig *et al.* [KIN 02] published in 2002, who suggested solutions for optical wireless communication devices, we can now glimpse totally secure wireless optical communication, from "n" objects to "m" objects and very high data rates (up to THz soon?) limiting itself to the walls of a room.

Optical wireless telecommunications also allow absolute security in communications, subject to having transmitters in a single reliable and reproducible photon. These free-space quanta in free space will certainly find useful applications for those who want absolute security in their information exchange.

Pierre-Noël FAVENNEC
URSI-France
March 2012

Acronyms

A	Ampere
AAC	Automatic attenuation control
Ac	Area cover
ACG	Automatic control gain
AEL	Accessible emission limit
AIR	Advanced infrared
AP	Access point
APD	Avalanche photodiode
APPM	Amplitude pulse position modulation
ARIB	Association of Radio Industries and Businesses
ARP	Address resolution protocol
ASCII	American standard code for information interchange
ASK	Amplitude shift keying
ATM	Asynchronous transfer mode
ATPC	Automatic transmit power control
AWGN	Additive white Gaussian noise
BCH	Bose–Chaudhury–Hocquenghem code
BCJR	Bahl–Cocke–Jelinek–Raviv code
BER	Binary error rate
BPM	Beam propagation method in time domain
BT	British Telecom
CAO	Concentrateur amplificateur optique (Fireball)
BC	Conduction band
CC	Convolutional code
CCD	Charge coupled device
CCETT	Centre Commun d'Etudes de Télévision et de Télécommunications
CD	Compact disc
CDMA	Code division multiple access

CEI	Commission Electrotechnique Internationale
CEPT	Conférence Européenne des Postes et Télécommunications
CIR	Channel impulse response
CNES	Centre National d'Etudes Spatiales
CNET	Centre National d'Etudes des Télécommunications
COFDM	Coded orthogonal frequency division multiplex
CPG	Conference Preparatory Group
CQI	Color quality indication
CRC	Cyclic redundancy check
CSI	Channel state information
CSMA	Carrier sense multiple access
CSMA/CA	Carrier sense multiple access with collision avoidance
CSMA/CD	Carrier sense multiple access with collision detection
DARPA	Defense Advanced Research Projects Agency
DC	Direct current
DD	Direct detection
DFB	Distributed feedback
DIF	Diffusion
DIV	Divergence
DLR	Deutsch Land Radio: German Spatial agency
DMT	Discrete multitone modulation
DPIM	Digital pulse interval modulation
DPPM	Differential pulse position modulation
DRM	Digital Radio Mondiale
DSL	Digital subscriber line
DSSS	Direct sequence spread spectrum
DVD	Digital versatile disc
ECC	Error corrector code
ECMA	European Computer Manufacturers Association
EDFA	Erbium-doped fiber amplifiers
EDRS	European Data Relay Satellite
EEL	Edge emitting laser
EFIR	Extremely fast infrared communication
EHF	Extremely high frequency
EN	European Norm (Euronorm)
ERO	European Radiocommunication Office
ESA	European Space Agency
Ethernet	LAN packet protocol
FCS	Frame check sequence
FDD	Frequency division duplex
FDDI	Fiber distributed data interface
FDMA	Frequency division multiple access
FDTD	Finite difference time domain

FET	Field effect transistor
FFT	Fast Fourier transform
FIR	Fast infrared
FOV	Field of view
FSO	Free-space optic
FTTx	Fiber to the Home, Business....
FTTH	Fiber to the home
GSM	Global system for mobile communications
GUI	Graphical user interface
HAP	High-altitude platform
HF	High frequency
HHH	Hirt–Hassner–Heise code
HP	Optical transmitted half-power angle
HTTP	Hypertext transfer protocol
IBM	International Business Machines
ICSA	Infrared Communication Systems Association
ICT	Information and Communication Technologies
I_d	Dark current
IdP	Indoor positioning
IEC	International Electrotechnical Commission
IEE	Institution of Electrical Engineers
IEEE	Institute of Electrical and Electronics Engineers
IIS	Interference intersymbol
IM	Intensity modulation
IM/DD	Intensity modulation/direct detection
InGaAs	Indium gallium arsenide
IP	Internet protocol
IPv6	Internet protocol version 6
IR	Infrared
IRC	Infrared communication
IrDA	Infrared Data Association
IrLAP	Infrared link access protocol
IrLMP	Infrared link management protocol
ISCA	Infrared communication Systems Association
ISI	Intersymbol interference
ISO	International Standards for Organization
ITS	Intelligent transport system
ITU	International Telecommunication Union
ITU-R	International Telecommunication Union Radiocommunication sector
JVC	Japan Victor Company
KDDI	Japanese telecommunication operator
LAP	Link access protocol
LASER	Light amplification by stimulated emission of radiation

LD	Laser diode
LCD	Liquid crystal display
LCR	Line clock recovery
LDPC	Low-density parity check code
LED	Light-emitting diode
LEOT	Laser electro-optics technology
LLC	Logical link control
LMP	Link management protocol
LOS	Line of sight
LRC	Longitudinal redundancy check
MAC	Medium access control
MIMO	Multiple-input multiple-output
MPDU	MAC protocol data unit
MPE	Maximum permissible exposure
MPEG	Moving Picture Experts Group
MRR	Modulating retroreflector
MS	Multispot
MSD	Multispot diffuse
MSDU	MSMAC service data unit
MSM	Metal-semiconductor-metal *photodiode*
Mozilla	Code name for the web Netscape Navigator
NASA	National Aeronautical and Space Administration
NEC	Nippon Electric Company Limited
NFIRE	Near-field infrared experiment
NLOS	Non-line of sight
NRZ	Non-return to zero
NTT	Nippon Telegraph and Telephone Corporation
OBEX	Object Exchange (IrDA exchange protocol)
OFDM	Orthogonal frequency division multiplex
OMEGA	HOME Gigabit Access
OOK	On–off keying
OPPM	Overlap pulse position modulation
OQAM	Offset quadrature amplitude modulation
OSI	Open systems interconnection
OWMAC	Optical wireless media access control
PC	Personal computer
PER	Packet error rate
PD	Photodiode
PDA	Personal digital assistant
PDU	Protocol data unit
PHY	OSI physical layer
PIN	Positive intrinsic negative diode
PLC	Power line communication

PLCP	Physical layer convergence procedure
PmP	Point-to-multipoint communication
PtP	Point-to-point communication
PPDU	PLCP protocol data unit
PPM	Pulse position modulation
PSDU	Physical service data unit
PSK	Phase-shift keying
QAM	Quadrature amplitude modulation
QKD	Quantum key distribution
QOFI	Qualité Optique sans Fil Indoor
QOS	Quality of service
RC5	Philips IRDA remote control protocol
RGB	Red green blue
RLL	Run length limited encoding
RR	Radio regulation
RS	Reed–Salomon code
RSA	Rivest–Shamir–Adleman code
RSV	Association of Reed–Salomon and Viterbi code
RS232	Universal data interface
RTSP	Real-time streaming protocol
RVR	Runway visual range
SAP	Service access point
SDMA	Space division multiple access
SEI	Space Exploration Initiative
SFD	Start frame delimiter
SFTF	Spaceborne flight test system
SHF	Super high frequency
SILEX	Semiconductor intersatellite link experiment
SIMO	Single-input multiple-output
SIR	Serial infrared
SIRSC	Sony IrDA data transmission protocol
SISO	Single-input single-output
SMTP	Simple mail transfer protocol
SNR	Signal-to-noise ratio
SPIE	Society of Photo-optical Instrumentation Engineers
SWO	Smart wireless optic
TIA	Transimpedance amplifier
TFTP	Trivial file transfer protocol
TCP	Transmission control protocol
TCP/IP	Transmission control protocol/internet protocol
TDD	Time division duplex
TDMA	Time division multiple access
TG	Task group

UDP	User datagram protocol
UFIR	Ultrafast infrared
UHF	Ultrahigh frequency
USB	Universal serial bus
UV	Ultraviolet
VB	Valence band
VCSEL	Vertical external-cavity surface-emitting laser
VFIR	Very fast infrared
VISPLAN	Infrared wireless LAN systems: WLAN system which combine IR technology (Ethernet 100 Mbps) and LAN mobility
VLC	Visible light communication
VLCC	Visible Light Communication Consortium
VoIP	Voice over IP
VRC	Vertical redundancy check
W	Watt
WDAN	Wireless domestic area networks
WDD	Wavelength division duplex
WDM	Wavelength division multiplexing
WDMA	Wavelength division multiple access
WIFI	Wireless communication protocols governed by IEEE 802.11 norms
WLAN	Wireless local area networks
WPAN	Wireless personal area networks
WLOS	Wide line of sight
WON	Wireless optical network
WS	Weapons system
WWRF	Wireless World Research Forum
WWW	World wide web

Introduction

Telecom operators are finding themselves confronted by a growing demand for a higher volume of information to be transmitted (voice, data, pictures, etc.). The increasing frequency in the systems used is a solution because it is able to offer higher bandwidth and allow higher flow rates. In the field of wireless communications, the use of links in the range of optical wavelengths, visible, ultraviolet, and infrared constitutes a form of wireless transmission of a few kilobits per second to hundreds of gigabits per second. They can be implemented either over short distances, limited to one room (office, living room, car, airplane cabin, etc.), or over medium distances (a few tens of meters to several kilometers) outside (atmospheric optical links or free-space optics – FSO), or over large distances in space (high-altitude platform – HAP, planes, drones, intersatellite, etc.).

This technique is not new. Over thousands of years, well before the work of the Abbot Claude Chappe, communication processes, although very primitive, were implementing optical transmission. But the amount of information provided remained low. Optical communications over long distances did not really start until the late 18th Century with the optical telegraph. But the quality of service (QoS) was low; the transmitters and receivers, men and materials' lack of reproducibility and reliability; and the transmission medium, the air, was changeable.

Soon, electricity (electrical charges) and copper replaced the optical (photons) and air. Transporting information through a copper line allows relatively high flow rates. At the beginning of the third millennium, these connections with copper as the medium are still widely used. For very large distances, for many decades, copper was the base material; it has covered the planet with a vast network of information transmission.

The invention of the laser in 1960 paved the way for an alternative solution – that of fiber optic telecommunication – offering a virtually unlimited transmission capacity. In 1970–1971, the almost simultaneous development of low-loss fiber optics and a semiconductor laser emitting in continuous operation at room temperature led to the explosion in wire optical communication. Glass is the medium for transmission of photons, and glass fibers may have lengths of several thousand kilometers. The optical wires were, therefore, unchallenged in underwater transmissions, transmissions over long distances, and interurban transmissions. It is the essential element of the information superhighway.

Since the liberalization of the telecommunications sector, motivation for the transmission of digital signals by the laser beam in free space is apparent. Several factors condition the renewal of this technology. First, regulatory reasons: there is no need for frequency authorizations or a special license to operate such links, in contrast to a large number of radio links. Second, economic reasons: the deployment of a wireless link is easier, faster, and less expensive for an operator than the engineering of optical cables. Finally, in the race for speed, the optical flow is the winner over the radio (even for millimeter wave) for desirable rates of several gigabits per second. In addition, the availability of components (lasers, receivers, modulators, etc.) widely used in optical fiber telecommunications technology potentially reduces equipment costs. The global market for digital wireless data transmission today is based primarily on radio wireless technologies. However, they have limitations and cannot be absorbed on their own, with a limited spectral width; development increases the need for higher speed.

The main applications of optical wireless focus on wireless telephony, information networks, and high-definition TV.

The objective of this book is to present the FSO that is currently used for the exchange of information, but, because of its many benefits (speed rates, low cost, mobility equipment, safety, etc.), it will explode as a telecommunications technique over the next decade and even become indispensable in computer architectures on short-, medium-, and long-range telecommunications.

From a didactic point of view, the book is organized into 12 chapters supplemented by two Appendices.

Chapter 1 discusses the basic concepts relating to light: the symbolism of the history, the different theories (wave, particle), the propagation and its various laws (reflection, transmission, refraction, diffusion, diffraction, etc.), interference, speed, spectral composition, emission, etc. That ends in 1960 with the laser invention, which opened up the way for many applications: CD, DVD, printers, computer disks, optical fibers, welding, surgery, etc.

Chapter 2, after some definitions related to telecommunications, reviews the various phases of the development of wireless optical communications over the centuries (smoke signals, light signals, movement of torches, etc.). And then in the 18th Century, after many tests, we review the appearance of Chappe's optical telegraph, the solar telegraph or heliograph, and the photophone of Graham Bell. Their principles (mechanism, code, etc.) are detailed and applications are described.

Chapter 3 presents the contemporary and the everyday life of wireless optical communications: the basic principles, the elements of electromagnetism, the electromagnetic spectrum, the propagation modes (line of sight, wide line of sight, diffusion, etc.), the different layers of OSI model, and the standardization aspects (VLC, IEEE 802.15.7, ECMA, IrDA). Then, contemporary and daily applications of wireless optical communication are described: indoor (limited space), outdoor (free-space optic), or spatial (links to aircraft, drones, HAP, intersatellite communications, etc.).

Chapter 4 is dedicated to the modeling of the propagation channel. It outlines the optical channel baseband and different types of modulation (on-off key (OOK), intensity modulation (IM), pulse position modulation (PPM), etc.). A comparison of the radio model is presented. The noise disturbance (thermal noise, periodic noise (artificial light), shot noise, etc.) is described. The signal-to-noise ratio compares the performance of different systems based on different technologies of digital communication. The channel is multipath (direct, reflected, diffused, etc.); the different paths are combined together. Intersymbol interference may occur. The different models of reflection (specular and diffuse (Lambert, Phong)) are presented. Reflection occurs when the wave encounters a surface on which the dimensions are large compared to the wavelength (floor, wall, ceiling, furniture, etc.). The reflection characteristics depend on the material surface, the wavelength, and the angle of incidence. Emphasis is then placed on the different models of diffusion.

Chapter 5 deals with propagation in the atmosphere. Atmospheric effects on propagation such as absorption and diffusion (molecular and aerosol particles), the scintillations due to the change in the index of air under the influence of temperature variations, and attenuation by hydrometeors (rain, snow) and their different models (Kruse, Kim, Bataille, Al Nabulsi, Carbonneau, etc.) are presented along with experimental results. The experiment implemented to characterize the channel optical propagation in the presence of various weather conditions (rain, hail, snow, fog, mist, etc.) is presented. Fog, whose presence is most detrimental to optical and infrared wave propagation, is explained (definition, formation, characteristics, and development). Visibility, the parameter that characterizes the opacity of the atmosphere, is defined. Measuring instruments for this characterization are described (transmissometer, scatterometer). The features of the "FSO Prediction" software simulating an atmospheric optical link in terms of probability of availability or interruption are described. It is a tool designed to help support decisions for the development of atmospheric optical links at high speeds over point-to-point links on short and medium distances.

Chapter 6 discusses the optical link budget in limited space, which is an important step in establishing a link. Knowing the sensitivity of the receiver, the goal is to calculate the power to implement at the emitter, to enable taking into account the losses in the optical channel. These various losses are identified and evaluated: geometric loss, optical loss, pointing loss, molecular loss, etc. Different cases are considered: a line of sight system and an optical system with reflection. The knowledge of the signal-to-noise ratio is then used to determine the error rate. It is connected to the different attenuations or disruptions of the transmitted signal in the channel.

Chapter 7 deals with immunity and standards' aspects as well as security and energy issues. For safety reasons, care must be taken to transmit power. Standards were developed by the International Electrotechnical Commission. They list the optical sources in seven different classes according to their level of dangerousness. Communication security is provided either in hardware or in software (encryption). The energy consumption of systems is an important parameter in choosing a technology. Finally, a presentation of the legislative aspect ends this chapter.

Chapter 8 entitled "Optics and Optronics" addresses the analog physical part of an optical device. Optical devices for transmission and reception and optical filtering are presented. The issue of optronics is then developed: the operating principle of the device and optronics emitters (white LEDs, infrared LEDs, laser, etc.) and receivers (photovoltaic cell, PIN photodiode, avalanche photodiode (APD), MSM photodiode, etc.).

Chapter 9 deals with data processing before the digital/analog conversion at the emission and after the analog/digital conversion at the reception. The data processing includes operations such as filtering, compression, analysis, prediction, modulation, and coding. Only modulation and coding parts in a specific configuration to optical wireless are described. Other items not directly related to the optical wireless are described elsewhere in the literature. Different modulations are explored: OOK, NRZ, ASK, QAM, PPM. OFDM and MIMO techniques are discussed. Coding aspects are detailed: principle, definition, performance, and many examples are mentioned: parity checks, cyclic redundancy check, block codes, BCH, RS, convolutional, etc.

Chapter 10 presents the "data link" layer, the second layer of the OSI system. The protocols of this layer handle service requests from the network layer and perform a solicitation of requests for services to the physical layer (downlink direction) and vice versa (upward direction). Access methods (TDMA, FDMA, CDMA, CSMA, WDMA, and SDMA) are described. The QoS parameters are mentioned. The various protocols used in wireless optical communications are presented for different types of data links: point-to-point (remote control, IrDA, VLC), point-to-multipoint (IEEE 802.11 IR, IEEE 802.3 Ethernet (ISCA-STB50), IEEE 802.15.3, IEEE 802.15.7, OWMAC).

Chapter 11 is dedicated to engineering of the installation of wireless optical communication in free space and limited space. In the area of free space (FSO), first there is a description of the principles of operation before turning to the characteristics of the equipment and recommendations for implementation. Optical budget calculations are detailed and examples of the availability of links for various French cities are presented. In the area of limited space, the habitat structure is first described: the distribution of areas of different rooms and the population percentage of a communication covered area. In the architecture of a wireless optical system, there is at least one optical wireless transmission/reception system per room, called base station (BS).

Each BS communicates with the terminals present in every room via a wireless optical communication. Finally, these terminals are connected or integrated to multimedia communication equipment (PC, monitor, PDA, etc.). Different simulations of optical system installations are carried out with a free software tool called "QOFI" and the link budget prepared: the base station is located in the middle of the ceiling (case A), above the door (case B), or on a socket (telephone, Ethernet, PLC (case C)); the terminal is installed in the lower opposite corner of the room (case 1), at a height equivalent to the top of a door (loud speaker, motion detector) (case 2), or on the ground in the middle of the room (case 3).

The aspects of the system are then discussed (the production of optoelectronics modules suitable for optical wireless, taking into account the safety aspect by using a diffuser at the emitter, obtaining an optical gain reception by setting in place an optical device called "fisheye", or processes such as equalization and OFDM, etc.).

Chapter 12 discusses the future of wireless optical communications in free and limited space at a home or an office. In each case, the advantages of this medium are underlined. The home and office potential are evaluated and faced with the economic and commercial realities.

Appendices remind the reader of various concepts related to optical geometric (refractive index, Snell's law, sources definition, image, focus, etc.), photometry (steradian, solid angle, etc.), and energy (light intensity, luminous flux, illuminance,

luminance, energy flow, lighting, geometric extent, etc.), and various items relating to the use of logarithmic notation (dB, dBW, dBm, etc.).

Various elements described in this book contributed to the development of new recommendations at ITU-R, the Radiocommunication Sector of the International Telecommunication Union, dedicated to propagation data and prediction methods required for the design of terrestrial free-space optical links and the definition of associated systems.

Chapter 1

Light

In the beginning God created the heavens and the earth. The earth was formless and empty, darkness was over the surface of the abyss, and the spirit of God was hovering over the waters. God said 'Let there be light' and there was light. God saw that the light was good: and God divided the light from the darkness. God called the light Day, and the darkness he called Night. And there was evening and there was morning, it was the first day.

> *"Fiat Lux – Let there be light"*
> Old Testament,
> The Pentateuch – Genesis 1,
> Chapter 1

Light has long fascinated man, exalted depictions by painters or praise from writers, with many areas of study for scientists and scholars. Figure 1.1 represents, for example, Lady Taperet (22nd Dynasty, 10th or 9th Century BC) praying to the sun god Ra-Horakhty. The symbolism of light provides an almost unlimited field for celebration of all kinds in all civilizations, past and present.

For centuries, the only known radiation was light. The first written analysis of light seems to date from Greek and Latin civilizations. For the Greeks, Euclid (325–265 BC) and Ptolemy (90–168 BC), the light is emitted from our eye and is the vector of an object image. On the other hand, Epicurus (341–270 BC) and the Latin poet Lucretius (98–55 BC) thought that the bright objects sent little pictures of themselves into space, referred to as "simulacras". These simulacras were entering our eyes so we could "see" these objects. This latter theory called "corpuscular theory of light" would be taken up again in a more abstract manner during the 17th and 18th Centuries.

Figure 1.1. *Stele of the Lady Taperet (Louvre museum)*

Because of this, from the 17th Century, the nature of light was a source of debate that lasted for more than 300 years. With the fundamental question, "Is light a wave or a stream of particles?"

To explain the laws of reflection and refraction of light rays, Rene Descartes (1596–1650) evokes particles that bounce off a mirror like a ball in a French game (*jeu de paume*) whose speed changes when entering a transparent medium (water or glass, for example). It is the source of the fundamental Snell–Descartes' laws. The authorship of the refraction law is attributed to Willebrord Snell (1580–1626) after Christian Huygens (1629–1695) refers to the date of the unpublished work of Snell on the subject. Note that the paternity of the discovery of the law of refraction is currently attributed to Ibn Sahl (940–1000) in 985. Ibn Al-Haytham (965–1039) wrote a book on optics (Opticae thesaurus) in which he mentions the phenomenon of refraction, but he could not develop the mathematical law. This discipline was originally called "dioptric", but later it was called geometrical optics for (or *due to the fact that*) the trajectory of light rays is built to geometrical rules.

Only a few decades later, Isaac Newtown (1643–1727) developed his particle model of light in 1704. It has a light composed of small "particles" emitted by luminous bodies moving very fast in a vacuum and in different transparent media. He does not hesitate to complicate the model to make it compatible with observations such as "Newton's rings". This interference phenomenon (Figure 1.2) is achieved by placing a lens (L) on a flat surface (P) with a light source (L'). It is possible to observe a series of concentric rings (A), alternating light and dark [NEW 18]. This is now explained by the wave approach.

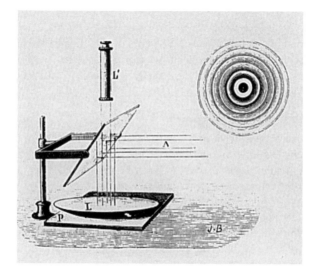

Figure 1.2. *Device and Newton's rings*

During the same period, Christian Huygens developed a wave model of light, by analogy with the wave propagation on the surface of the water. This model also explains the phenomena of reflection and refraction. But, with his particular prestige acquired by his law of universal gravitation, Newton turned off the debate, and imposed his corpuscular theory of light onto the scientific community at the time.

It was not until about a century later that the existence of many known phenomena was explained by geometrical optics (decomposition of light, interference, etc.) returning to the wave approach with studies of Thomas Young (1773–1829) and Augustin Fresnel (1788–1827). The "wave theory of light" defines the light as a vibration, similar to sound, vibrating in an invisible environment called "Ether".

Because measurements were not possible with the instruments of the time, an initial estimate of the propagation speed was 200,000–300,000 km/s with a very important frequency of vibration. This model is predominant when explaining the phenomena of interference and diffraction.

Finally, almost half a century later, James Clerk Maxwell (1831–1879) offered four fundamental equations that summarized the knowledge of the time in the electrical, magnetic, and electromagnetic fields. He succeeded in electromagnetic fields by applying what Newton had done in the field of mechanics. One of these, the Maxwell–Ampere equation defines light as an electromagnetic wave consisting of electrical fields and magnetic fields vibrating transversely with a velocity of 300,000 km/s.

This is the electromagnetic wave theory of light and this model, faced with measures of speed of light, dedicates Maxwell's proposal. But visible light from red to violet is a special case of those electromagnetic radiations, as Maxwell predicted the existence of other radiation emissions from natural or artificial sources (e.g. cosmic rays or radio transmitters).

In fact, in 1887 Heinrich Hertz (1857–1894) invented an electromagnetic wave transmitter whose frequency is below infrared frequencies (below the red). These frequencies, known as radio frequencies, are the wave bands of radio and television. Then in 1895, Wilhelm Röntgen (1845–1923) discovered very high frequency radiation higher than the ultraviolet frequencies, this is X-rays.

In 1900, Max Planck (1858–1947) made a significant contribution, with the explanation of the spectral composition (color distribution) of emitted light and the quantification of energy exchange between light and matter. These energy exchanges are realized by integer multiples of an indivisible base quantity (Figure 1.3). These quanta or quantum of energy are related to a given frequency radiation multiplied by a constant. This new constant of physics is called Planck's constant (h) and is initiated by quantum physics.

A few years later, in 1905, Albert Einstein (1879–1955) hypothesized that light was made up of energy (photons) and he proposed a corpuscular theory of light. The laws of Fresnel and Maxwell are still valid, but the energy approach shows that the same wave transports energy called photons. This last point helps to explain such phenomena as the photoelectric effect (discovered by Hertz in 1887). And in 1909, despite reticence from the scientific world at that time to reconcile his theory with the electromagnetic wave model, Einstein concluded that light is both a wave and a particle.

Then, from the theories of Rutherford (1871–1937), Neils Bohr (1885–1962) takes the opposite approach in 1913 and published a model of atomic structure and chemical liaison. This approach became a model (the Bohr's model), the atom has a nucleus around which gravitate electrons. The farthest orbits from the core comprise more electrons, which determine the chemical properties of the atom. These electrons move from one layer to another by emitting or absorbing a quantum of energy, the photon. Figure 1.3 shows the emission of a photon by de-excitation of an electron (left case) and the excitation of an electron by absorption of a photon (right case). Planck's constant (h) relates the energy E (E = Ea − Eb) of a photon or an electron (e) to its frequency by the relation $E = h\nu$.

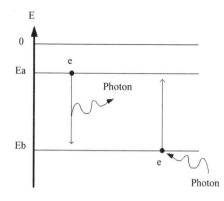

Figure 1.3. *Emission and absorption of a photon*

In 1924, Louis de Broglie (1892–1987) generalized the wave–particle duality of matter, by proving that an electron can also be a wave.

With Leon Brillouin (1889–1969), Erwin Schrödinger (1887–1961), Werner Heisenberg (1901–1976), and Max Born (1882–1970), demonstrated that reality consists of elementary particles that are either in the wave form or in the corpuscular form. The light is a manifestation wave (electromagnetic approach) photon (particle approach). These revolutionary advances occurred at scientific conferences, including the International Congress of Physics of Solvay in Brussels in 1927. This particular conference was attended by the following figures as shown in Figure 1.4. In the first row, from left to right: Max Planck (second), Marie Curie (third), Hendrik Antoon Lorentz (fourth), Albert Einstein (fifth), Paul Langevin (sixth); the second row: William Lawrence Bragg (third), Paul Dirac (fifth), Louis de Broglie (seventh), Max Born (eighth), Niels Bohr (ninth); in the third row: Erwin Schrödinger (sixth), Wolfgang Pauli (eighth), Werner Heisenberg (ninth), and Leon Brillouin (eleventh).

These were the first steps of wave mechanics and then, modern quantum physics.

Figure 1.4. *The International Congress of Physics, Solvay in Brussels (1927)*

Even today, light has not fully disclosed all of its mysteries, there will be new discoveries. For example, recent studies have revealed the existence of "slow light", offering potential for the realization of quantum computer memories or the possibility of "squeezed states", reduction of certain quantum fluctuations of light.

But the best known example of the "use" of light is the laser. The main feature of the LASER (light amplification by stimulated emission of radiation) is to obtain homogeneous photons. These photons have the same propagation direction, the same frequency, and the same polarization. The first laser (1960) had pulses of the order of microseconds (10^{-6} s) and now we can obtain a level of the femtosecond (10^{-15} s), which allows the detection of electronic motion in matter. Among the countless applications of lasers [BES 10], the correction of myopic eyes to optical scanning devices, is worth mentioning as an application in the field of communications. This is the fiber optic communication, without which society and information communication technology (ICT) would not have become so important. Another discipline, the subject of this book, is wireless optical communications.

Chapter 2

History of Optical Telecommunications

"I have heard a ray of the sun laugh and cough and sing"
1884, Alexander Graham Bell,
Inventor of the photophone and probably the phone,
Patent 235.496 from 12/14/1880

2.1. Some definitions

2.1.1. *Communicate*

The verb "communicate" appeared in the French language around 1370. It was derived from the Latin *communicare* meaning "to share, import". The idea became enriched as a result of the meaning of the Latin *cum* (with) and *municus* (burden).

With this idea of sharing, the word first got a sense of participation in something. It lost the right to be in mutual relationship, in communion with someone. From the 16th Century, the word experienced a new extension in its meaning "transmit" (communicate news, spread disease, share a sense). It is also used in physics, "communicate the heat".

2.1.2. *Telecommunication*

The definition of the word "telecommunication" adopted during the 1947 International Radiotelegraphic Conference held at Atlantic City (USA) is:

any transmission, emission or reception of signs, signals, writings, images, sounds or information of any nature by wire, radioelectricity, optics or other electromagnetic systems.

The means of transmission must be of the electromagnetic type, which gives a very wide scope, since, as Maxwell showed, electromagnetic waves include electricity and optics.

2.1.3. *Optical telecommunication*

This involves any transmission, emission, or reception of visual signs and optical signals on guided and unguided support.

2.1.4. *Radio frequency or Hertzian waves*

These are electromagnetic waves of frequency lower than 3,000 GHz; they are propagated in space without artificial guidance (in optics, frequencies are significantly higher: hundreds of terahertz).

2.2. The prehistory of telecommunications

In the beginning, the need to remotely communicate was a natural reaction to life in a community. Communication would have been essential from the earliest times: even from the days of Adam and Eve. The first men were already "wireless operators", since they communicated with each other without wire, using sound and luminous waves (optical signals).

In ancient times, signal fires were used extensively to communicate and transmit a message from one place to another. The smoke was used during the day and the fire was used during the night.

Homer mentions light signals in the *Iliad* (late 8th Century BC): the fall of Troy was announced by fires lit on the hilltops. The *Anabasis* indicates this same mode of correspondence between Perseus and the army of Xerxes in Greece. In *Agamemnon*, Clytemnestra says to Coryphee agreeing with Agamemnon that the announcement of the capture of Troy reached him quickly through relay fires. As soon as Ilion was taken, a fire shone on Ida, a fire that other places copied, to transmit the news. In total, nine bonfires were used probably covering about 550 km relay: Ida, the rock of Hermes in Lemnos, the Mount of Athos, Makistos in Euboea, Messapios, rock of Cithaeron, Epiglancte, the Arachne mount, and Mycenae.

In 426 BC, Thucydides (460–400/395 BC) reports that during the Peloponnesian attack on Corcyra, signs of fire were brandished during the night, announcing the arrival of 60 Athenian ships. Based on this warning signal, the Peloponnesian fleet left quickly during the night.

Later, the Greeks developed more complex processes operating between cities or from one island to another. Aeneas the Tactician (4th Century BC) conceived a code of communication based on fire and water (Figure 2.1). To send a message, two groups of two men on each side had an earthenware pot of exactly the same diameter and height. The base of these vases is a hole of exactly the same size. Initially, the hole is plugged and the two containers are filled with a meter of water. In each of the vases, a cork floats and serves as the cap, with a rod, along its entire length having a succession of military terms placed at a distance determined by each other. These are military terms relating to possible events in war, e.g. riders arrived in the country, heavy infantry, and ships.

Figure 2.1. *Aeneas the Tactician device*

When one of the events listed on the stick occurs, simple torch signals suffice for groups to come into contact. Once both groups have established contact, they hide the torches and open the holes of their containers.

The corks are lowered at the same time as the liquid drains away, and when the relevant military term of one of two groups reached the top of the container, the operator shakes the torch again to show the flow is 'off'. Recipients only have to read the words inscribed on their rod and so on for the next message. Polybius (210/202–126 BC) improved this system by replacing monoalphabetic symbol, by multiple representation, the Polybius square or square of 25. This square is based on a system for converting letters into movements of torches. The letters of the alphabet are divided into five groups of five letters arranged in columns and rows. And to convey a message, it was enough to raise torches on the left to indicate the column and torches on the right to indicate the row.

This transmission process evolved very quickly in a cryptographic technique. Indeed, with such a simple letter shift, even though an enemy could observe the signals exchanged, he/she could not know its meaning.

Later, the Carthaginians linked Africa to Sicily by very bright signals. Then the Romans appropriated the formula and communication throughout the Roman Empire was driven by fire signal lines located on watchtowers that Caesar made great use of in his campaigns in Western Europe.

Fires located on high points along the great Roman roads were used to transmit quickly essential military information of a rudimentary character (Figure 2.2). With the help of such fires, lit from hill to hill, the Roman general Aetius sent to Rome the news of his victory over Attila in AD 451.

Figure 2.2. *Roman telegraph station (low-relief of the Trojan column in Roma)*

In the same way, ancient Chinese texts discuss the use of fires from watchtowers and even kites, during the Warring States (4th Century BC). This particular mode of communication, from different lamps lit on kites, was primarily military. This method was used until the period of the siege of Nanking (1937). Among North American Indians, the information was transmitted by smoke signals.

For a long time, sailors used semaphores and hand signals. As mentioned before, the idea of alphabetic signs dates back to Greek antiquity, but it was not until the Middle Ages that the first semaphore signals were used.

These methods of communication, although very primitive, were implementing optical means of transmission. The speed of transmission was already adequate, but the amount of information contained in each message was extremely low because the different possible configurations of light sources were very limited.

2.3. The optical aerial telegraph

Although the prehistory of telecommunication extends over millennia, the history of telecommunications really only starts at the end of the 18th Century with the appearance, in France, of Claude Chappe's (1763–1805) optical telegraph.

True semaphore, formed using wooden arms in a code corresponding to letters of the alphabet, seems to have been invented in 1684 by the Englishman Robert Hooke (1635–1703). His machine (Figure 2.3) consisted of a broad screen that revealed letters or signals, which corresponded to a code that had been agreed in advance. The system was improved in 1690 by the French physicist Guillaume Amontons (1663–1705). He suggested the use of a telescope to read the signals formed on the screen from a greater distance, in order to establish a correspondence, and thus a communication, between two distant points.

Figure 2.3. *Robert Hooke's telegraph*

The machine consists of a large screen (width, height) seen from a distance. Various signals can be seen. They indicate either the alphabet or phrases agreed in advance.

During the whole of the 18th Century, proposals for semaphore telegraphs followed one after another: Dupuis, Linguet, Courréjoles, and Brestrasser (arm signals) drew up various projects. However, it was the system of Abbot Claude Chappe that was adopted by a convention in 1793, on Joseph Lakanal's recommendation. In the ordinance dated September 24, 1793, Chappe was given permission to use towers and bell towers to install telegraph equipment between Paris and Lille. This connection, 230 km long, was built in record time between 1793 and 1794.

Chappe's telegraph machine was related, in principle, to semaphore. At the top of the towers (Figure 2.4), in line of sight, was a system built of three articulated arms and driven by a mechanical device (Figure 2.5): each of the arms could take various positions. The set of the possible combinations constituted a code that could be deciphered remotely because of the use of telescopes: the distance between the relay stations was, at most, only a few tens of kilometers.

Figure 2.4. *The Chappe's system used high points (towers, bell towers, etc.)*

The principal branch AB, horizontal in the figure, was approximately 4 m long and two small branches called wings, AC and BC, were approximately 1 m long.

The mechanism is under the roof of the tower or bell tower on which mast DD is fixed. The mobile branches are cut out in the shape of shutters, which give the properties of great lightness, and wind resistance. The mobile branches are painted black to ensure good visibility against the background sky. Branch AB can give four positions: vertical, horizontal, right to left, and left to right diagonal lines. Wings AC and BC can form right, acute, or obtuse angles with branch AB. The figures thus formed are the signals. They are clear, easy to see, easy to write, and it is impossible to confuse them. They allow a significant number of configurations.

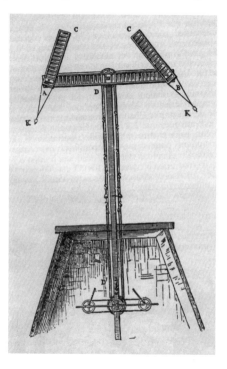

Figure 2.5. *Mechanism developed by Chappe: three articulated arms*

The first great success of Chappe's telegraph was the announcement to the French Convention of the victory at Landrecies (July 19, 1794) and then the resumption of the attack on Conde and on the Austrians by the troops of the Republic (September 1, 1794).

Chappe was appointed Director of Telegraphs to reward this success. His invention was called the optical telegraph. This involved transmission over a really long distance so this system represented considerable progress compared to the other media existing at the time.

Thus, the Chappe's telegraph entered history. However, it would only really take its place a few years later when the second link was built on the strategic Paris–Strasbourg axis. This link, which included 50 stations, was brought into service in 1798, and the first dispatch was transmitted on July 1, 1798. It announced the capture of Malta by General Napoleon Bonaparte, during his Mediterranean Sea crossing toward Egypt. Next, a link was installed between Paris and Brest (1798), then another toward the south of France (1799). The Paris–Lyon link was installed in 1806 and was extended toward Turin and Milan (1809). The northern link reached Antwerp, via Brussels, the same year.

A few decades later, the Louis-Philippe government undertook the construction of a network designed on the pattern of a spider's web, with Paris at its center in order to improve the efficiency of its administration. This network supplemented the previously installed links between Paris and the larger French cities (Lille, Brest, and Strasbourg), then installed concentric links, centered on Paris, recreating the radial links from place to place. Such a network, (Ethernet premise), offered the ability to benefit from an extraordinary escape route for the transmission of news if communications were cut off from Paris. After this, the network was enlarged by linking the northern fortified towns, the commercial areas around the coast, and the major cities of the south of France. There were even several clandestine links, including the Paris–Lyon link conducted in 1836 by businessmen who wanted to benefit from a communications network. Thus, in 1844, France had a network of 534 semaphore stations covering more than 5,000 km. This Chappe's "air" network had a common characteristic with all the French communication networks, in particular roads and railways: it was built on a star topology centered on Paris.

The Chappe's invention was also very much exploited in other countries, especially in Spain and Italy. In Russia, Tsar Nicholas I established links between Moscow, St Petersburg, Warsaw, and Cronstadt. He inaugurated the Moscow–Warsaw links in 1838. These links comprised 200 stations served by 1,320 operators. In the most difficult areas, fog made the lengthened signals hard to see. The system was adapted to make use of mobile shutters giving combinations that were varied enough to offer a multitude of signals. Free-space telegraphs were also built according to this system in England and Sweden.

2.4. The code

To have an operating telegraph, Chappe had to solve two problems: the mechanism and the code. To be effective, the code was not alphabetical: the transmission of the shortest sentence would have required as many signals as letters, which would have taken a considerable time. Thus, Chappe adopted a set of codes: signals or rather groups of signals were linked to words, sentences, or geographical names (Figure 2.6). Each signal was made using three articulated elements. A characteristic geometrical figure was created by varying the angle of each element. These shapes could be identified from a distance (Figures 2.7 and 2.8). The central arm was either horizontal or vertical (Figure 2.7(a)), then to it is added a left or right diagonal (Figure 2.7(b)). The two wings could be horizontal, vertical, or left or right diagonal.

Let us delay a little on this invention which introduced the concept of the information and telecommunications network. A link was composed of a succession of towers, at the top of which a mechanism formed of articulated arms was installed whose various positions, 92 in all, each corresponded to a number. Ethnologists have since highlighted that ants exchange information thanks to this type of signal. In each relay station tower, an operator playing the role of repeater-regenerator reproduced, without understanding the significance, the code dispatched to him by the preceding tower, so that it could be observed with a telescope. The operator of the next station transcribed the message received on paper, in the form of a succession of numbers, and carried it to the addressee who, using a highly confidential coding document, possessed by only a few, associated a word to each number and thus reconstituted the message. This system simultaneously combined digital transmission; the encoding intended to protect information; and error correction. The numerical data rate was low since it took approximately 20 seconds to transmit a combination, which corresponds to approximately 0.4 bit/s, but the Chappe's optical telegraph brought an extraordinary gain to the transmission speed compared to the best method known hitherto, the transport of the mail by a rider. The first message was transmitted by this technique on March 2nd 1791 and a veritable network was constituted little by little: it would continue to be used until around 1855, when it would be supplanted by the electrical telegraph, which offered the advantages of a higher transmission capacity and 24/7 usage, including during the night as well as during times of low visibility.

Thus the term "optical" disappeared for a time from the transmissions field.

Figure 2.6. *The Chappe's telegraph [JOI 96]*

The code of the Chappe's telegraph evolved over the course of time. However, it should be noted that, from the beginning, the Chappe's telegraph did not use alpha-digital code, but a code similar to that of diplomats: the idea came from a relative of the inventor Leon Delaunay, former French Consul in Lisbon. The signs transmitted using the arms of the apparatus corresponded to figures or simple numbers. At the beginning, the number of signals was limited to 10; the articulated arms could only take horizontal or vertical positions (see Figure 2.7(a)). Then, in about 1800, the number increased from 10 to 88, then to 92. The first code used by Chappe was composed of 9,999 words, each one represented by a number. Later, there were three codes: a "word" code, a "sentence" code, and a "geographical" code. These three codes were amalgamated into one code in 1830: this new code is undoubtedly that which is preserved at the Post Office Museum at Paris.

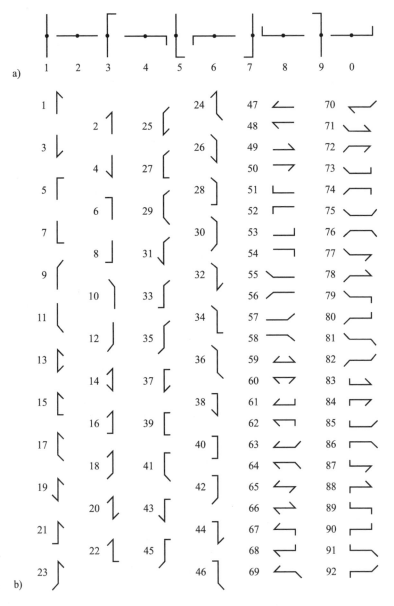

Figure 2.7. *Signals used in the final code of Chappe's telegraph*

The "vocabulary" of this code is contained in an imposing register of 704 pages (88 × 4 + 88 × 4), comprising some 45000 terms, sentences or geographical names. In theory, the transmission was done by group of two numbers; the first corresponded to the vocabulary page and the second to the text line.

Thus, one could have, for example, under the "army" heading:
*page 36, line 6: "**army**"*
page 35, line 9: "enemy army composed of"
page 35, line 13: "the army completely defeated the enemy"

*and, under the heading "**message**":*
page 53, line 12: "message of utmost importance"
page 53, line 13: "The end of this message I am not yet reached"
page 53, line 21: "I am responding to your last message"

Thus a text such as:
"I am responding to your last message, the army completely defeated the enemy" required only the two groups: 53–21 and 35–13

Such a system ensured the secrecy of the messages and economy of signals at the same time. It also explains the stereotyped style of the dispatches sent by this means during the time of the French Revolution and the French Empire.

Figure 2.8. *The Chappe's code [Libois, unpublished]*

The development of electricity led to the birth of the electric telegraph. In 1832, and from the work of Andre-Marie Ampere (1775–1836) and François Arago (1786–1853), Samuel Morse (1791–1872) invented a simple and robust data transmission system and patented the electric telegraph in 1838.

This system consists of a power line between two points with a transmitter and a receiver positioned at each end. On transmission, a manual switch supplies, with a battery, more or less briefly the line. At the reception, an electromagnet connected directly to the line activates a mechanism to translate the code marked on a paper strip, which advances at a rate of pulses on the line.

The code invented by Samuel Morse is a transcript of a series of dots and dashes of the letters of the alphabet, numbers, and common punctuation. The point is a short pulse and the line is a long pulse. For example, SOS, an acronym defined *a posteriori* as Save Our Souls or Save Our Ship results in three short pulses followed by three long and three short pulses (... --- ...).

In 1838, the first electric telegraph worked between London and Birmingham, and after 1845 it gradually takes the place of the optical air telegraph. Indeed, the electric telegraph offered a much higher transmission capacity and, especially, permanent use. In France, the last line of the air telegraph network was withdrawn in 1859.

2.5. The optical telegraph

However, the optical air telegraph improves and becomes a tool oriented to new uses, especially in special circumstances, such as transmissions between army corps during war or peacetime. Based on Chappe's telegraph, the optical telegraph produced visible signals at great distances, by means of light and mirrors. The main change is the creation of these signals corresponding to characters of the Morse code. They consist of flashes or eclipses of sunlight or an oil lamp.

2.5.1. *The heliograph or solar telegraph*

The heliograph or solar telegraph is a portable optical telegraphy system able to transmit signals over distances of 40 km. Here, we use the reflection of sunlight off a bearing mirror. The reflected rays are directed toward the station in correspondence. Unlike other devices previously seen, based on controlled light signal emission, the telegraph transmits signals by luminous flux extinction (at rest, it still illuminates the station in correspondence). From the first English version (Figure 2.9), Leseurre improved the system that was used especially during the siege of Paris in 1870–1871. While the electric telegraph systems had been ransacked, and therefore, become ineffective, it was due to the optical transmission system (Figure 2.10) that correspondence could still be sent to the forts around Paris. France was then happy to have a communication system that passed over the heads of the enemy. But the major disadvantage is that is functions only during the day because the sun is the only power for transmission.

The operation of the English heliograph is as follows. When we wish to transmit signals, the mirror M is oriented and the reflected rays will cross the target D; this system works well only if the sun is high above the horizon (Max Nansouty, "electricity", 1911).

Regarding the Lesuerre's heliograph, a second mirror is used to collect the sun's rays falling on the first one with a very important angle and to reflect them on the main mirror. The double mirror system can be used on a time range much wider than the previous system (CNAM, Nature, Tissandier, 1881).

Figure 2.9. *An English heliograph*

Figure 2.10. *Lesuerre's heliograph*

2.5.2. *The night and day optical telegraph*

Colonel Mangin proposed a significantly improved system. The source is an oil lamp that, by a set of mirrors and shutters, gives a parallel flashing light beam. The more or less prolonged brightness reproduces the dots and dashes of Morse's telegraphy. A diffracting telescope attached to this set, parallel to the axis of the lenses, can see the signals of the corresponding post. During the day, the oil lamp can be replaced by an optical system recovering sunlight (Figure 2.11).

In Figure 2.11, left, the light source is an oil lamp and shutter A controls the transmitting light signals; the diffracting telescope EE' parallel to the axis of the lenses is used to see the signals from the corresponding post. On the right, the telegraph for use during the day is illuminated by sunlight, the two plane mirrors M and M' suitably inclined in accordance with the position of the sun return the light beam refracted into the device, and through a set of lenses we have then an output light beam (Max Nansouty, "electricity", 1911).

Figure 2.11. *Mangin's telegraph*

The speed of the emission of light flashes is about 20 words per minute. With sunlight during the day and that of an oil lamp during the night, the lens optical devices can communicate up to distances of 30–120 km. The signals are always seen through a diffracting telescope attached to the receiving device (Figure 2.12: CNAM, Nature, Tissandier, 1881). These great distances necessarily require absolute clear weather because the light signals can be interrupted by smoke and fog, however minor.

2.6. Alexander Graham Bell's photophone

Therein the first electronic version appears in 1880. Four years after he invented the telephone, Alexander Graham Bell (1847–1922) filed with Charles Sumner Tainter (1854–1940) the description of a device called the photophone.

The operating principle of the photophone is revolutionary in the use of electricity (Figure 2.13). Sunlight is focused on a reflective flexible membrane. The user speaks into the diaphragm of this membrane. The speech is transmitted in the air by modulating the reflected sunlight. This modulated reflected light, after moving in the air, is collected by a photoconductive selenium cell connected to a battery and headphones.

Figure 2.12. *Using the optical telegraph of Colonel Mangin*

Figure 2.13. *The photophone*

In the figure (Figure 2.13), left, sunlight is modulated by a vibrating diaphragm and then transmitted in free space on a distance of about 200 m. Right, this beam is picked up by a selenium cell itself connected to headphones.

"I have heard a ray of sun laugh and cough and sing", he wrote to his father after an initial demonstration of operation. It was a free-space transmission over a distance of 600 ft (about 183 m) transmission using the sunlight as the carrier. Although it is the phone wire that history has preserved for posterity, Bell has always considered the photophone as his greatest invention: "The greatest invention I have ever made, greater than the telephone".

The photophone has failed to win favour with the wire phone and to get a place in the world of telecommunications because of too low transmission distances and the lack of light sources that emit continuously. This invention, which never became a commercial reality, however, demonstrated and highlighted the basic principle of modern optical communications, where optical fiber has replaced the atmosphere.

In addition, and a few years after the discovery of Hertz (see Chapter 1), a radical change came upon the field of wireless telegraphy. From 1890, many scientists and inventors have shown the relevance and increase in distance of a wireless radio transmission: Nikola Tesla (1856–1943) for energy conversion; Alexander Popov (1859–1906), for the demonstration of a wireless radio receiver; Albert Turpain (1867–1952) in 1895, by sending and receiving up to 25 meters of a radio signal in Morse code; and Guglielmo Marconi (1874–1937), by sending and receiving of a radio signal up to 6 km in 1896. Figure 2.14 shows a photograph of Marconi radio equipment taken at the *Museum of the History of Science* in Oxford.

Figure 2.14. *Marconi radio equipment*

The letter "S" transmission by Marconi in 1901 over the Atlantic Ocean led to the use of optical wireless devices in the background.

But less than a century later and with the invention of the laser, engineers returned to the ideas of the photophone and think of using wireless optical links for wireless digital transmissions at very high speed. It will not be the sun that will be used but perfectly controllable light beams.

These photon beams are provided by lasers whose characteristics are known (wavelength, the number of photons emitted in time or in emission sequence, etc.) and are then collected by highly sensitive and reproducible sensors. The transmission medium is air and can be changed for external communications (climate, vegetation, industrial, etc.) like for radioelectricity and broadcasting communications.

During the last 30 years, the technology of telecommunications systems based on wireless optical links has changed significantly and has developed mainly in two practical areas: in outdoor environments with, for example, within defense and the aerospace engineering field, and in indoor environments with a well-known example, the remote control of home appliances.

Chapter 3

The Contemporary and the Everyday Life
of Wireless Optical Communication

*"2010 is the beginning of the decade of
wireless optical communications"*
Dr Larry B. Stotts,
Defense Advanced Research
Projects Agency (DARPA),
IEEE Spie Optics & Photonics,
San Diego (USA), August 2009

3.1. Basic principles

The wireless optical communication systems were the first to be used (Figure 3.1) and, due to technological innovations, they become attractive, mainly because of the lack of produced or suffered radio interference and, especially, from the ever-increasing demand of the data rate.

Before presenting the daily and contemporary applications of wireless optical communication, we define some basic principles and the different modes of propagation.

3.1.1. *Operating principle*

3.1.1.1. *Block diagram*

Wireless optical communications refer to the use of light propagation in the field of optics with free space as a transmission medium.

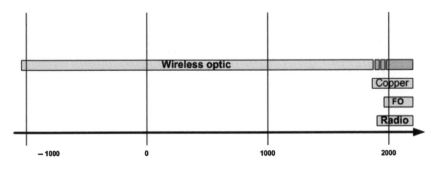

Figure 3.1. *Communication history*

An example of the principles of operation is detailed in Figure 3.2 with various technical modules present in an optical wireless in emission (Tx) and in reception (Rx). At the transmission (Tx), from digital data received by a computer or a server via a suitable interface (D-Tx), a digital and analog electronic processing (El-Tx) is performed to adapt the signals to electrical-optical converter (Ot-Tx). This converter is usually a light-emitting diode (LED) or a laser. Using the divergence optical device (Op-Tx), a beam is emitted more or less divergent to the receiving device. At the reception (Rx), part of the received beam is optically concentrated (Op-Rx) to be focused on the optical-electrical converter (Ot-Rx), which is a photodiode. Electronic processing (El-Rx) is inversely made to adapt the signals to the interface (D-Rx). This communication is also possible in the opposite direction and provides a two-way communication.

Figure 3.2. *Modules of an optical wireless device*

In most cases, a monitoring software is supplied with devices. This software allows us to configure the connection and obtain qualitative and quantitative information of the various modules.

One of the main parameters in the characterization of a wireless optical device is the distance: it varies depending on the equipment from a few tens of centimeters (IrDA equipment) to several tens of thousands of kilometers (intersatellite link). The rate and type of application recommended is also essential. The proposed data rate starts from a few kilobits per second or Kbps (103 bit/s) to several terabits per second or Tbps (1012 bit/s). The different applications are discussed later in this chapter.

3.1.2. *The optical propagation*

There are several propagation characteristics in optical wireless links. The propagation between the emitting source and the receiving cell can be carried out in three systems (hybrid systems combine one and/or the other of these characteristics):

– line of sight (LOS);

– wide line of sight (WLOS);

– propagation by diffusion with specular and diffuse reflection (DIF) also sometimes referred to as non-line of sight (NLOS).

3.1.2.1. *Line of sight propagation – LOS*

This is the simplest link and best-known system for point-to-point (PtP) communication. In this case, the transmitter and the receiver are manually or automatically pointed (tracking), one toward the other to establish a link (Figure 3.3).

The LOS from the transmitter to the receiver must be clear of obstructions and the majority of the transmitted beam is directed toward the receiver. This is the configuration of the majority of outdoor wireless optical communication systems (free-space optic – FSO) and optical communication systems between computers or devices (IrDA). These links can be created temporarily for a session of data exchange between two users (or computer systems), or permanently set in a local network.

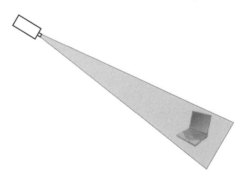

Figure 3.3. *Line of sight*

This type of propagation is characterized by:

– the absence of any obstacle in the space between the transmitter and the receiver;

– high power efficiency, due to the low divergence emission (a few degrees). Much of the signal is routed to the receiver;

– a less important receiver sensitivity to ambient light because field of view (FOV) of the receiver is lower;

– better resistance to multipath and, by extension, to intersymbol interference (ISI), again because of a low FOV of the receiver;

– an available throughput only function of the link budget and not from the dispersion of multipath and problems related to ISI, the latter is similar to an echo phenomenon.

These LOS systems, therefore, have very limited flexibility and high sensitivity to the obstacles and are particularly suited to a PtP configuration.

Unfortunately, these narrow beams do not provide a large coverage for communication. To remedy this problem, two approaches are possible and are described below.

In the first approach, a manual or automatic pointing device [MAM 98] is installed, the response times are fast, but the costs are currently incompatible with the creation of a common network.

In the second approach, a multiple transmitting/receiving device with multiple coverage angles is used [PAR 01]. In this multisectoral or cellular approach (Figure 3.4(a)), it is possible to ensure high coverage while maintaining high data rates [WIS 97].

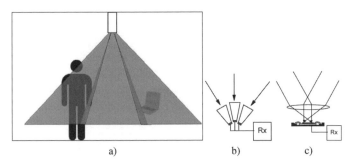

Figure 3.4. *(a) Multisectoral, (b) angular diversity, and (c) imaging reception*

The implementation of multiple transmitter and receiver couples, each couple having a specific and different angle of coverage, can be achieved by the introduction at the reception of an angular diversity [CAR 00, OME 09], presented in Figure 3.4(b). Another more compact approach [YUN 92] is shown in Figure 3.4(c) for the reception part, for example. It consists of several receptors (matrix array) arranged on a flat surface and associated with a converging lens [KAH 98]. This device allows an imaging reception where the detector will be different depending on the angle of the incident beam [OBR 03]. This second solution offers a significant miniaturization of optronics modules.

3.1.2.2. *Wide line of sight – WLOS*

Configuration for WLOS (Figure 3.5) is characterized by an emitter with a larger divergence angle, so more coverage, and receivers with a wide angle of reception. This extended coverage makes it easier to overcome an alignment between the transmitter and the receiver, but the downside is that only a small fraction of the emitted power is received by the receiver. So there is a significant degradation of the link budget.

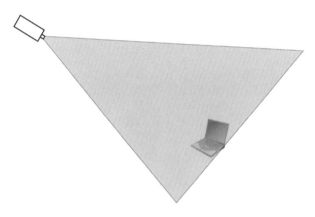

Figure 3.5. *Wide line of sight*

In addition, the implementation of the high divergence emitter leads to the appearance of reflections on walls and objects in a room. This results in the possible appearance of multiple paths arriving on large FOV receptors. From a certain distance between the transmitter and the receiver, it is a side effect that restricts the data rate and the quality of the transmission; this is known as ISI.

The ISI is an unwanted phenomenon for a wireless optical communication, which is manifested as a signal distortion. Indeed, the symbol transmitted at time T-1 and having bounced off a wall, for example, can become a noise and affect the transmitted symbol at time *T*. This is equivalent to an echo phenomenon and makes communication less reliable. Orthogonal frequency-division multiplexing (OFDM) and discrete multitone (DMT) modulation are solutions in the field of data processing that reduce the ISI (see Chapter 9).

WLOS links are more adapted to point-multipoint (PmP) communication. A typical example is an access point that allows for communication with receiving devices in the covered area [SMY 95] and the best-known product is the remote control device.

3.1.2.3. *Diffusion propagation (DIF) and controlled diffusion*

In a diffuse system, the connection is always maintained between any transmitter and any receiver in the vicinity, by reflecting the optical beam on surfaces such as ceilings, walls, and furniture (Figure 3.6). In this case, the transmitter and the receiver are not necessarily directed toward each other, the emitter uses a highly divergent beam, and the receiver has a large FOV, so that the direct path is not required.

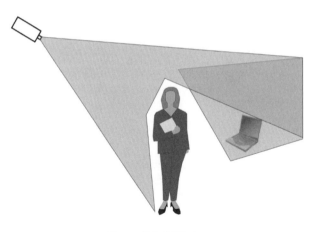

Figure 3.6. *Diffusion*

Such a mode of operation ensures a communication, regardless of the obstacles or the person preventing the direct path. However, the penalty of the diffuse transmission is that the capacity is reduced compared to the LOS system. This is mainly due to losses that are typically over 100 dB and a multibeam receiver, which create ISI.

The signal-processing solutions are required to reduce this phenomenon, but an approach known as controlled diffusion can also solve this problem.

The first controlled diffusion device was proposed in the late 1970s (Figure 3.7) by Gfeller and Bapst [GFE 79] at a rate of 1 Mbps. Then a data rate of 50 Mbps was reported in the 1990s by Marsh and Kahn [MAR 96]. This system illuminates the walls and ceiling with controlled lighting, coupled with a receiver that accepts the reflected beam in a limited angular aperture. It aims to combine the robustness of the diffuse system with high data rates.

Many publications have proposed such geometries [DJA 00], including the use of transmitters that illuminate small areas on the ceiling [YUN 92]. They are used in conjunction with a narrow FOV receiver to capture the different incident beams with similar optical path.

The latter two regimes are well suited to the application of wireless local area networks such as WLAN (wireless local area network) and WDAN (wireless domestic area network) because users do not need to know the location and orientation of the various communicating devices.

Figure 3.7. *Controlled diffusion*

3.1.3. *Elements of electromagnetics*

The transmission of information can take different forms. The use of transfer of information from the information that can convey the electromagnetic waves is one of the richest possibilities. More details about electromagnetic waves are discussed in the next section. This section will consist of a classic reminder of electromagnetic waves. We discuss the Maxwell equations that control the propagation of these waves in different environments, then we recall the expression of the energy carried by these waves, and, finally, we present the propagation rays models.

3.1.3.1. *Maxwell's equations in an unspecified medium*

The propagation of electromagnetic waves is conducted by Maxwell's equations that are [BRU 92, COZ 83, VAS 80]:

$$\overrightarrow{\text{rot}}\,\vec{E} = -\frac{\partial \vec{B}}{\partial t}$$

$$\overrightarrow{\text{rot}}\,\vec{H} = \vec{j} + \frac{\partial \vec{D}}{\partial t}$$

$$\text{div}\,\vec{B} = 0$$

$$\text{div}\,\vec{D} = \rho$$

where

– $\vec{E}$ is the electric field (V/m).

– $\vec{H}$ is the magnetic field (A/m) associated with the electromagnetic wave.

– $\vec{D}$ and $\vec{B}$ are, respectively, electric (C/m^2) and magnetic (Wb/m^2 or T) induction, which describe the response of the medium where the electromagnetic wave propagates.

– ρ is the electric charge density (C/m^3).

– $\vec{j}$ is the current density (A/m^2). $\vec{j}$ and ρ are linked by the conservation of charge relation (equation [3.2]).

Equation 3.1. *Maxwell's equations (unspecified medium)*

$$\text{div}\,\vec{j} + \frac{\partial \rho}{\partial t} = 0$$

Equation 3.2. *Conservation of charge relation*

These equations allow us to treat the propagation of electromagnetic waves in any medium. If we consider the propagation in free space which is the subject of this book, the previous set of equations is simplified. Indeed, the wave propagation takes place in a homogeneous environment and without charge ($\rho = 0$ and $\vec{j} = \vec{0}$).

In addition, we consider the transmission medium as isotropic, and linear or non-dispersive:

$$\vec{D} = \varepsilon \vec{E}$$

$$\vec{B} = \mu \vec{H}$$

where ε and μ are, respectively, the permittivity or dielectric constant and the permeability or magnetic constant.

Equation 3.3. *Equations for a free-space propagation*

The simplest case is vacuum propagation where $\varepsilon = \varepsilon_0 = 1/(36\pi 10^{-9})$ (F/m) and $\mu = \mu_0 = 4\pi 10^{-7}$ (H/m). In a specific medium, we will call $\varepsilon/\varepsilon_0$ the relative permittivity ε_r and μ/μ_0 the relative permeability μ_r.

To address this problem, we consider the case of sinusoidal time. In other words, we look after solutions that exist for $t = -\infty$ to $+\infty$ and assume that the fields and inductions vary sinusoidally with time and at the same pulsation ω. For example, the real electric field denoted by $\vec{E}_r$ can be written as:

$$\vec{E}_r = \vec{E}_0 \cos(\omega t + \varphi)$$

Equation 3.4. *Electric field (1st equation)*

To simplify the calculations, it is generally preferred to use the following complex notation:

$$\vec{E}_c = \vec{E}_0 \exp(i\varphi)$$

$\vec{E}_r$ is obtained from $\vec{E}_c$ by:

$$\vec{E}_r = \text{Re}\left\{\vec{E}_c \exp(i\omega t)\right\} = \frac{1}{2}\left\{\vec{E}_c \exp(i\omega t) + \vec{E}_c^* \exp(-i\omega t)\right\}$$

where $\vec{E}_c^*$ is the conjugate complex.

Equation 3.5. *Electric field (2nd equation)*

In the rest of this section, we use the complex notation, the field will simply be noted $\vec{E}$.

3.1.3.2. Propagation of electromagnetic waves in an isotropic medium

The goal of this part is to characterize the waves propagating in a linear and isotropic homogeneous medium. The medium is homogeneous if the permittivity ε is constant. Let us consider the simplest case, i.e. the propagation in a homogeneous medium without any electrical charge. Using the complex notation, this can be written as:

$$\overrightarrow{\text{rot}}\,\vec{E} = -i\omega\mu\vec{H}$$

$$\overrightarrow{\text{rot}}\,\vec{H} = i\omega\varepsilon\vec{E}$$

$$\text{div}\,\vec{H} = 0$$

$$\text{div}\,\vec{E} = 0$$

Equation 3.6. *Maxwell equations (isotropic and linear homogeneous medium)*

From the four preceding relations, the two following equations are deduced:

$$\Delta\vec{E} + \omega^2\varepsilon\mu_0\vec{E} = \vec{0}$$

$$\Delta\vec{H} + \omega^2\varepsilon\mu_0\vec{H} = \vec{0}$$

where Δ is the Laplacian operator.

Equation 3.7. *Propagation equations*

These two relations are generally known as the equations of propagation. Their analytical resolution shows that:

$$\vec{E} = \vec{E}_0\exp(-i\vec{k}\cdot\vec{r})$$

$$\vec{H} = \vec{H}_0\exp(-i\vec{k}\cdot\vec{r})$$

where $\left|\vec{k}\right|^2 = k^2 = \omega^2\varepsilon\mu_0$

Equation 3.8. *Electric and magnetic field*

$\vec{k}$ is the wave vector or propagation vector. It expresses the spatial frequency of the wave. The spatial period λ is linked to $\vec{k}$ through $\left|\vec{k}\right| = \frac{2\pi}{\lambda}$.

$\vec{E}$, $\vec{k}$, and $\vec{H}$ form a right-handed orthogonal trihedral of vectors. In other words, if the electric field $\vec{E}$ propagates along the direction given by its wave vector $\vec{k}$, the associated magnetic field $\vec{H}$ has a single direction. The real fields $\vec{E}_r$ and $\vec{H}_r$ vary sinusoidally with time. Their dependence on the distance is shown in Figure 2.1. For a propagation in a homogeneous, dielectric, and isotropic material, $\vec{E}_r$ and $\vec{H}_r$, as well as their associated complex notations, lie in a plane normal to the direction of propagation. In this case, the electromagnetic vibration is said to be transverse.

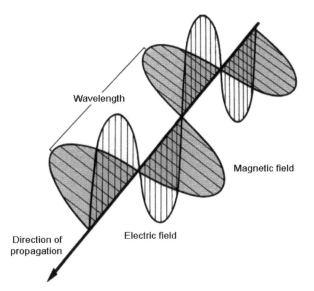

Figure 3.8. *Representation of the real components of the electric and magnetic fields along the direction of propagation*

From the previous relations, the phase of the wave can be written in the form:

$$\varphi(\vec{k}, \vec{r}) = \omega t - \vec{k} \cdot \vec{r}$$

Equation 3.9. *Phase of the wave*

The phase velocity, i.e. the speed necessary for remaining at a constant phase, is:

$$v_\varphi = \frac{\omega}{k}$$

Equation 3.10. *Phase velocity*

For a wave propagating in a medium:

$$v_\varphi = \frac{c}{n}$$

where

    – c is the velocity of light in vacuum.

    – n is the refractive index of the medium.

Equation 3.11. *Phase velocity in the vacuum*

3.1.3.3. *Energy associated to a wave*

The energy balance is one of the most significant parameters for the link. Until now, we have considered plane waves extending in all the planes normal to the direction of propagation. For optical links, the beams will be limited laterally. We shall recall here the expression for luminous flux through surfaces in the case of plane waves.

To calculate the energy carried by a wave, let us first recall that the electromagnetic energy present in a volume V is expressed by the relation [BRU 92]:

$$U = \int_V (\varepsilon |\vec{E}|^2 + \mu_0 |\vec{H}|^2) dV$$

Equation 3.12. *Electromagnetic energy in a volume V*

The energy density is given by:

$$w_{em} = \frac{1}{4}(\varepsilon |\vec{E}|^2 + \mu_0 |\vec{H}|^2)$$

Equation 3.13. *Energy density*

In the case of a plane wave propagating in a perfect medium, it can be shown that the electrostatic energy is equal to the magnetic energy. The electromagnetic energy density is therefore:

$$w_{em} = \frac{1}{2}\varepsilon\left|\vec{E}\right|^2$$

Equation 3.14. *Electromagnetic energy density*

Let us now consider the luminous flux through an uncharged homogeneous surface, normal to the direction of propagation, where the normal to the surface is parallel to the z-axis. Let us consider the amount of energy during time $dt = dz/v$, where v is the speed of light in the medium considered. If the electric and magnetic fields are linearly polarized along Ox and Oy, it can be shown that the energy flux through the plane is equal to the flux of a normal vector $\vec{P}$ given by [BRU 92]:

$$P_z = \frac{dU}{dz} = \frac{1}{2}E_x H_y{}^*$$

where $H_y{}^*$ is the conjugate complex.

Equation 3.15. *Energy flux*

The Poynting vector is written as:

$$\vec{P} = \frac{1}{2}\vec{E}\wedge\vec{H}^*$$

Equation 3.16. *Poynting vector*

The instantaneous energy associated with the wave cannot be measured due to the fact that the frequency of light is about 10^{15} Hz, and the detectors measure only the average energy. Indeed, only the energy average over a period T is observable:

$$\langle\vec{P}\rangle = \frac{1}{T}\int_0^T \vec{P}dt = \frac{1}{2}\mathrm{Re}(\vec{E}\wedge\vec{H}^*)$$

Equation 3.17. *Average energy value*

The average flux of the Poynting vector through a surface corresponds to the average energy carried by the wave through this surface.

NOTE. – Here again, it is necessary to be careful depending on whether the real or complex notation of the electromagnetic fields is used [VAS 80].

3.1.3.4. *Propagation of a wave in a non-homogeneous medium*

The plane waves described previously give the ideal case of an infinite, homogeneous, linear, and lossless medium. The case of wave propagating in air is close to this ideal case. Of course part of the electromagnetic energy propagating in the atmosphere will be absorbed by molecules in the air; the absorption process will be described in detail in Chapter 5. In a bounded medium, as is the case in an optical fiber or waveguide, it is necessary to take Maxwell's equations regarding the differences between the media into account. The equations of propagation will no longer assume the simple form given by equation [3.8]. Since a purely analytical solution is not always available, it is necessary to use numerical algorithms for the resolution of the propagation equations of an electromagnetic wave in the system under consideration. Let us mention some algorithms such as the *finite difference time-domain* (FDTD) method, the *beam propagation method in time domain* (BPM), and the finite element integration method.

3.1.3.5. *Coherent and incoherent waves*

In free-space propagation, the diffraction and reflection processes generate several waves that interfere with one another (Figure 3.9). If two waves have a phase relation, i.e. they are coherent, the fields will be summed in amplitude and phase. The resulting field depends on the phase difference between the waves. If the two waves are incoherent with respect to one another, i.e. if there is no phase relation, the intensities still sum up. Thus, it is important to consider the phase relation between the detected signals.

Figure 3.9 shows the pattern at the detector level where two light beams will be added, one that has spread in a direct line (a) and another that has reflected on a surface roughness (b), and 2θ is the FOV of the beam.

There is a more complete account of coherence and interference phenomena in the references given in the preceding section. Here, we shall only mention a few properties defined for plane waves, as well as some definitions and relations associated with the coherence of different detected signals.

It has been shown that the density of electromagnetic energy is proportional to the square of the electric field.

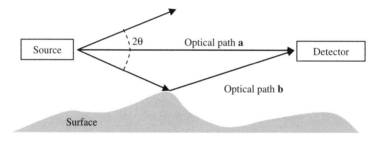

Figure 3.9. *Free-space propagation with reflection*

Let us consider the fields E_a and E_b emitted by the source after their propagation along the paths a and b with lengths L_a and L_b, respectively:

$$\vec{E}_{a,b} = \vec{E}_0 \exp(-jkl_{a,b})$$

The total field can be written in the form:

$$\vec{E}_{tot} = \vec{E}_0 \exp(-jkl_a) + \vec{E}_0 \exp(-jkl_b)$$

Equation 3.18. *Total field from two sources*

The average intensity of the field at the detector is therefore:

$$I = \left|\vec{E}\right|^2 = 2\left|\vec{E}_0\right|^2 [1 + \cos k(l_b - l_a)]$$

Equation 3.19. *Average intensity of the total field*

Variation in the signal can be observed due to the phase difference between the beams. The ideal case corresponds to that of perfectly monochromatic waves, where the wave emission of the source is infinite. Reality can be very different: the term including the phase difference can be zero. In this case, the two beams are said to be incoherent with respect to one another, and therefore their respective intensities will be directly added. The reduction in the coherence of a source can be associated either with the space incoherence of the source related to its spatial extension or with the spectral width of the source that is not null as in the case of an ideal monochromatic light.

When the length difference between the two optical paths varies, the energy passes through minima and maxima. The visibility V is defined by:

$$V = \frac{I_{Max} - I_{min}}{I_{Max} + I_{min}}$$

Equation 3.20. *Visibility from two optical paths*

The visibility depends on the degree of coherence of the source, on the length difference between the paths, as well as on the location of the detector with respect to the source. The coherence among various beams arriving at the detector also depends on the crossed media: for example, the diffusing medium can reduce the coherence. We can use coherent sources on condition that parasitic reflections are not interfering with the main beam, inducing modulations of interference signal. Lasers have a coherence length longer than a meter, whereas the coherence length of white light sources can be of the order of the micron.

3.1.3.6. *Relations between electromagnetism and geometrical optics*

In the course of this book, geometrical optics will be extensively used. First, we shall briefly outline the transition from electromagnetism to geometrical optics and explain why the use of ray optics is fully justified in this context. Let us consider the case of an actual non-homogeneous medium. The refractive index can vary. Thus, we search for solutions of the following form:

$$\vec{E} = \vec{E}_0(r)\exp[-ik_0 S(\vec{r})]$$
$$\vec{H} = \vec{H}_0(r)\exp[-ik_0 S(\vec{r})]$$

where $k_0 = \dfrac{2\pi}{\lambda} = \dfrac{\omega}{c}$ are related to the vacuum.

Equation 3.21. *Equations of an optical ray*

This wave is a harmonic wave. The quantity $S(\vec{r})$ is called *Eikonal*, $S(\vec{r})$ is comparable to a distance and corresponds to the optical path, while $\vec{E}_0(r)$ and $\vec{H}_0(r)$ are vector quantities. The readers interested in the *Eikonal* function are referred to the reference [COZ 83]. Here, we briefly describe some of their properties, as well as the relation between rays and waves.

Using Maxwell's equations, equation [3.1], and with the notation $\vec{k}(\vec{r}) = k_0 \overrightarrow{grad}(S)$ in the equations above, we obtain:

$$\vec{k} \wedge \vec{E}_0 - \omega\mu\vec{H}_0 = -i\overrightarrow{rot}\vec{E}_0$$
$$\vec{k} \wedge \vec{H}_0 - \omega\varepsilon\vec{E}_0 = -i\overrightarrow{rot}\vec{H}_0$$

Of course, if $\vec{E}_0(r)$ and $\vec{H}_0(r)$ were constant, we would find the same relations as in the case of a plane wave. The plane wave approximation is justified, provided that:

$$\frac{\Delta E}{E} << \frac{\Delta x}{\lambda_0}$$
$$\frac{\Delta H}{H} << \frac{\Delta x}{\lambda_0}$$

where ΔE and ΔH are the variations of the amplitudes of electric and magnetic fields when we move from x to $x + dx$.

Thus, the harmonic wave turns out to be similar to a plane wave. In other words, if the variations in amplitude of the fields are small with respect to the wavelength, the harmonic wave can be considered locally as a plane wave.

The wavefront is the locus of the points that are in phase, i.e. where $S(\vec{r})$ is constant. For a plane wave, the wave front is a plane. For a harmonic wave, it may assume any specific shape (see Figure 3.10).

In this figure, each surface element is equivalent to its tangent plane. The beam normal to the local plane sets out a beam of light. The normal to the wave front of the harmonic wave is given by the vector $\vec{n}_f$, which of course depends on the point considered in the wave front:

$$\vec{n}_f = \frac{\overrightarrow{grad}S}{|\overrightarrow{grad}S|}$$

Equation 3.22. *Vector normal to the wave front*

For any point, the vector $\overrightarrow{grad}S$ defines a normal to the wave front associated with this point. This corresponds to the definition of a light ray in geometrical optics. Likewise, $\overrightarrow{gradS}$ is related to the refractive index of the medium, where the harmonic wave propagates through the equation:

$$\left|\overrightarrow{gradS}\right|^2 = c^2\varepsilon\mu = n^2$$

where c is the velocity of light.

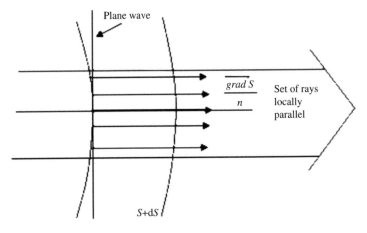

Figure 3.10. *Local approximation of a wave surface by a plane wave*

The concept of plane wave, like that of light ray, is a purely theoretical notion used for approaching reality in a simplified way. The concept of light ray assumes a point source coming from infinity, while the concept of plane wave supposes a plane wave extending to any point of space. This means that in the following, every time the notion of light ray is used, it is implicitly assumed that a plane wave can be defined locally.

The free-space loss is due to scattering or absorption of energy as the wave moves away from its emission source. It is a function of distance from the place and time of emission, the environment in which the wave propagates, and its frequency.

In the vacuum, electromagnetic waves propagate with a speed called speed c (299,792,458 m/s). Depending on the frequency or wavelength, we speak, for example, of radio waves, light waves, or X-ray.

The period of the wave is $T = 1/v$, its wavelength is $\lambda_0 = cT$ in the air, and $\lambda = cT/n$ in a medium of index n (the index of air is 1, and the water index is about 1.33).

This representation of the wave radiation is well suited to the study of phenomena such as diffraction, which involve radiation–radiation interaction. Regarding the study of radiation–matter interactions, for example when there is an exchange of energy status of optoelectronic components, the particle representation is better suited.

In this approach, it is defined as a plane wave frequency v is associated with a particle: the photon energy $E = hv$ and momentum $p = h/\lambda$, where h is Planck's constant ($h = 6.62606896 \times 10^{-34}$ J.s).

3.1.3.7. *The electromagnetic spectrum*

All these different frequencies or wavelengths form what is called the electromagnetic spectrum. It is divided into two categories: ionizing bands and non-ionizing bands, depending on the impact of these waves on biological tissues and matter. The ionizing part of the electromagnetic spectrum includes what is defined as alpha, beta, gamma, and X-rays. The wavelength is then very short, <60 nm, so frequency is very high. The non-ionizing radiation is defined by metrics waves, radio waves, microwaves, infrared, visible light, and ultraviolet.

This spectrum is very wide; it extends from a few hundred hertz to several billion hertz. Figure 3.11 shows the extent of the electromagnetic spectrum based on the frequency. There are successive bands of radio (Radio), infrared (IR), visible (Vis), ultraviolet (UV), X-rays (X), and gamma rays (γ).

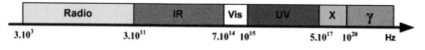

Figure 3.11. *Electromagnetic spectrum*

3.1.3.8. *Units and scales*

As can be seen, the spectrum of electromagnetic waves is very wide, with wavelengths ranging from 10^{-14} m up to 10^4 m. Both very long and very short wavelengths will be encountered throughout this book. Table 3.1 gives the meaning of some prefixes that are used on these occasions.

Prefix	Symbol	Unit	Nominal value	Numerical value
Exa-	X	10^{18}	One quintillion	× 1,000,000,000,000, 000,000
Peta-	P	10^{15}	One quadrillion	× 1,000,000,000,000,000
Tera-	T	10^{12}	One trillion	× 1,000,000,000,000
Giga-	G	10^{9}	One billion	× 1,000,000,000
Mega-	M	10^{6}	One million	× 1,000,000
Kilo-	k	10^{3}	One thousand	× 1,000
Milli-	m	10^{-3}	One-thousandth	× 0.001
Micro-	μ	10^{-6}	One-millionth	× 0.000001
Nano-	n	10^{-9}	One-billionth	× 0.000000001
Pico-	p	10^{-12}	One-trillionth	× 0.000000000001
femto-	f	10^{-15}	One-quadrillionth	× 0.000000000000001
Atto-	a	10^{-18}	One-quintillionth	× 0.000000000000000001

Table 3.1. *Prefixes commonly used in electromagnetism*

In the field of optical wireless communications, it is possible to define four spectral bands currently in use. These bands are shown in Figure 3.12 and ranged from 100 nm to 3,100 nm. For comparison, the spectrum of the Sun at sea level is also shown (black curve).

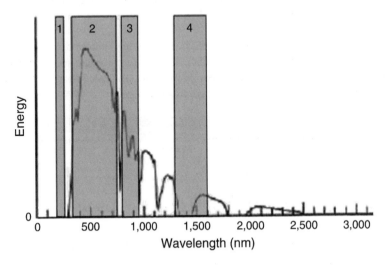

Figure 3.12. *Spectral band*

The four spectral ranges are:

– band from 200 nm to 280 nm or ultraviolet band or C-band;

– band from 350 nm to 750 nm or visible band (visible light communication – VLC);

– band from 800 nm to 950 nm or near-infrared band 1 (first window of fiber-optic communications and optical wireless);

– band from 1,300 nm to 1,600 nm or near-infrared band 2 or Telecom band (last window of fiber-optic and optical wireless communications).

Each band offers advantageous features and constraints consistent within the parameters outlined below:

– eye safety;

– optical disrupting;

– the optic link budget;

– ease of industrialization, that is to say, a low complexity;

– the transmitted power;

– the sensitivity;

– etc.

The choice of wavelength is therefore an important parameter in the realization of a wireless optical link.

We should also mention a few units used in the field of optical links:

– The amplitude of the electric field $\vec{E}$ is measured in volts per meter (V/m).

– The amplitude of the magnetic field $\vec{H}$ is measured in ampere per meter (A/m).

– The energy density of electromagnetic field is measured in watts per square meter (W/m^2), or in milliwatts per square meter (mW/m^2), or in milliwatts per square centimeter (mW/cm^2).

Fundamentally, a wave is an oscillation, i.e. a periodic variation of a physical state that propagates in space or through matter. It is characterized by its amplitude, its direction of propagation, its speed, and its frequency in hertz.

The number of oscillations per unit of time is referred to as the frequency, and is measured in hertz (Hz).

The time interval between two successive oscillations with same direction and size is called the period. Its unit is the second (s).

The space traveled in by the wave during this time is the wavelength. Its unit is the meter (m).

The set of points reached by the disturbance in a homogeneous medium after a given time, starting from the time of emission, is referred to as the wave surface or wave front.

The relation among these different units (frequency, period, and wavelength) is presented below:

– The frequency $f = 1/T$.

  - f = frequency in hertz (or multiple);

  - T = period in seconds (or multiple).

– The wavelength $\lambda = c \times T = c/f$.

  - T = period in seconds (or multiple);

  - c = velocity or speed of light = 3×10^8 m/s or 299,792 km/s;

  - f = frequency in hertz (or multiple).

– The period $T = 1/f$.

  - T = period in seconds (or multiple);

  - f = frequency in hertz (or multiple).

Depending on the field of study, different units are used for various spectral bands. Table 3.2 shows the relations among the frequency, the period, the wavelength, and the characteristics, depending on the part of the electromagnetic spectrum being considered. Different parameters are generally chosen merely for ease of use, but are associated with the same physical phenomenon.

However, it is recommended to use the international system MKSA (meter, kilogram, second, ampere) in calculations.

Waves	Wavelength	Frequency	Period	Characteristics
Radio waves	100 km–1 m	10 kHz– 300 MHz	100 ns–0.1 ms	Radio FM, AM, and TV
Microwaves	1 m–1 mm	300 MHz–300 GHz	3 ps–100 ns	Mobile, satellite, radar
Infrared	1 mm– 0.8 μm	300 GHz–300 THz	3 fs–3 ps	Laser, night sight, telemeter
Visible light	400–800 nm	Energy in electron-volts (eV) –1 to 5 eV	1–3 fs	Laser, Sun, lamps
Ultraviolet	400–0.5 nm	5–1 keV	1.7as to 1 fs	Laser, lamps
X-rays	50–0.1 pm	1–100 keV	0.0003–1.7 as	X-ray tubes
Gamma-rays	<0.1 pm	>100 keV	< 0.0003 as	Radiation of energetic particles, synchrotrons

Table 3.2. *Frequencies, periods, and corresponding wavelengths of different electromagnetic spectra*

3.1.3.9. *Examples of sources in the visible and near visible light*

We now present a few examples of optical sources and provide some information concerning solar radiation. Indeed, it is important to know the spectral signature of these sources of perturbation for optical links since these radiations may disturb optical links in free space.

The main natural source of electromagnetic radiation is the Sun. Natural electromagnetic energy (solar radiation) allows, among other things, photosynthesis of trees and plants. Its spectrum extends from 300 nm to more than 1,500 nm, with varying intensities or amplitudes. The peak intensity is located approximately 480 nm (corresponding to the color blue) before progressively decreasing as the wavelength increases. Our eye perceives but a small fraction of solar radiation, that between 400 nm and 700 nm (see Figure 3.13). The Sun emits incoherent radiation. The Sun's spectrum is shown in Figure 3.13 (for good visibility, the absorption peaks of oxygen molecules, ozone, and water are not shown) in intensity per unit frequency and MKSA unit is $Wm^{-2}Hz^{-1}$.

The Sun can be approximated as a black body. Another example of a black body is a filament lamp. The intensity of radiation from a black body at temperature T is given by Planck's law of black body radiation:

$$F(v) = \frac{8\pi h v}{c^3}\left[\exp\left(\frac{hv}{kT}\right) - 1\right]^{-1}$$

where

- T is the temperature of the element (in K);

- h is Planck's constant;

- k is the Boltzmann constant;

- c is the velocity of light;

- v is the frequency.

Equation 3.23. *Planck's law of the Sun*

For the Sun, this spectrum exhibits a maximum at the frequency $f = 4 \times 10^{14}$ Hz.

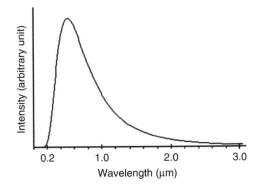

Figure 3.13. *Curve of brightness of solar radiation*

Another source of electromagnetic radiation, commonly used in both office and home environments, is the tungsten filament lamp.

Figure 3.14 shows the emission spectrum of a 60 W lamp. This spectrum is very broad and continuous, and exhibits a peak around 1,000 nm.

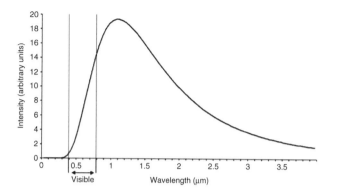

Figure 3.14. *The optical spectrum of a tungsten filament lamp*

It might be stressed that optical communications use sources with a quasi-monochromatic narrow bandwidth (Figure 3.15) whose spectra can be compared instructively with the spectra resulting from ambient lights (the Sun, lamps, etc.). However, it is clear that for optical communication in free space, any source emitting in the spectral window of the detector is likely to disturb the transfer of information.

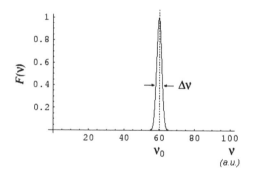

Figure 3.15. *Spectrum of quasi-monochromatic source (v_0)*

3.1.3.10. *Conclusion*

Since free-space telecommunication relies on the use of electromagnetic waves, it seemed essential to us to briefly describe some properties of these waves. Likewise, since geometrical optics is extensively employed for simulating free-space links, it was necessary to present the relation between electromagnetic waves and light rays, as well as the practical limitations of geometrical optics.

3.1.4. *Models for data exchange*

3.1.4.1. *The OSI model*

To describe the essential features for communication between computers in networks, the International Organization for Standardization (ISO), a dependent of the United Nation (UN) and made up of 140 national standards bodies, proposed a model called the Open Systems Interconnection (OSI) reference ISO 7498.

This model describes a particular reference level in network modeling. It consists of seven layers (see Figure 3.16):

– The physical layer transmits signals between equipment. The data are bits, for example RS232, xDSL, or WiFi.

– The data link layer manages communications or error-free transmissions of frames between two adjacent network equipment (MAC layer, such as ATM and Ethernet).

– The network layer manages communication step-by-step and provides the functions of addressing and routing packets (e.g. ARP and IPv6).

– The transport layer handles communications between end-to-end processes (running programs) and makes cutting or reassembly information (e.g. TCP and UDP).

– The session layer manages the synchronization of exchanges and allows opening or closing trading session (e.g. *Real Time Streaming Protocol* (RTSP) and Telnet).

– The presentation layer and code structure of data exchanged by applications (e.g. ASCII and Videotex).

– The application layer is the user access to network services via the application (e.g. HTTP, VoIP, and Mozilla).

Each level consists of the following:

– A service that provides primitives. These are commands, such as a connection request "connection.request", or events, such as receiving data "data.indication".

– A protocol that consists of a set of messages and rules for exchanges to achieve this service. The messages of a protocol are called *protocol data unit* (PDU). Some features of a protocol, such as the detection of transmission errors, the correction of these errors, and flow control, may be present in several levels.

– An interface that is defined as the access point to service in the standard. It is characterized by the library functions in a program, for example, or by a set of registers to the input of a circuit in a hardware configuration.

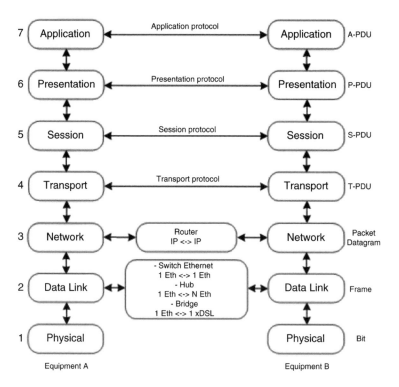

Figure 3.16. *OSI model example*

The process of data exchange between layers is generally similar. After a connection request from the applicant "connection-request", the requested informs the demand "connection-indication" and responds either positively or negatively "connection-response", which is validated by the applicant "connection-confirmation". The latter event is an acknowledgment. Data transfer follows the same process ("DATA.request", "DATA.indication", and "DATA.confirm") [TOU 03].

For example, to emit, the layer level N + 1 sends data to layer level N with the primitive "DATA.request". These data will be encapsulated according to the protocol layer N + 1 (N + 1-PDU). In turn, the layer of level N will encapsulate the data according to its own protocol (N-PDU) before transmitting to N − 1. Upon receipt, each layer protocol analysis corresponding to its layer de-encapsulates the data and transmits them to the top layer with the primitive "DATA.indication".

But the text of the standard is very abstract because there was the desire to make it applicable to virtually any network. As a result, a different model, simpler and more widely used, has emerged: the Transmission Control Protocol/Internet Protocol (TCP/IP) model (Figure 3.17).

3.1.4.2. *The DoD model*

The TCP/IP model was developed in the mid-1970s as part of the research DARPA (Defense Advanced Research Projects Agency, – the USA). This model should meet the needs of interconnects computer systems of the Army (Department of Defense – DoD).

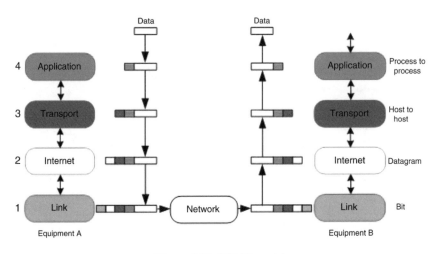

Figure 3.17. *TCP/IP model*

It incorporates the modular approach, but it consists of four layers:

– The network layer includes the physical layer and data link layer of the OSI model. For example, Ethernet is a typical implementation of this layer because it can send and receive IP packets in a network.

– The Internet layer performs the interconnection of remote networks. It provides the functions of routing packets independently of each other until the final destination. The routing function allows us to compare it to the network layer of the OSI model. Its name is Internet Protocol (IP).

– The transport layer is the same as that of the OSI model. In this model, this layer has two implementations: TCP and User Datagram Protocol (UDP); it can transmit data faster but less reliable because there is no acknowledgment.

– The application layer, unlike the OSI model, is immediately above the transport layer and contains all higher level protocols such as Telnet, Trivial File Transfer Protocol (TFTP), Simple Mail Transfer Protocol (SMTP), and HyperText Transfer Protocol (HTTP).

3.2. Wireless optical communication

It was the U.S. Department of Defense (DoD) that was the first to recognize the value of optical wireless links because of their potential to provide links that could not be intercepted or interrupted. From 1944, the Navy tested an infrared military telephone. The phone operated with an infrared lamp using vapors of cesium (852 nm and 894 nm) for transmission and a photocell associated with an audio amplifier for reception.

Subsequently, by studying the main technological challenges of engineering related to these telecommunication systems, the activity of defense and aerospace engineering has helped establish a strong technical and scientific basis on which were based commercial optical communication systems.

Then, in the late 1990s with the advent of the Internet and the strong need for speed in the telecommunications sector, several companies have developed a new generation of outdoor optical communication systems for commercial use and adapted to the private sector. Then in the mid-2000s, faced wih the growing need for home or business use, and to deal with the various proposed radio solutions with the constraints of power and spectrum resource, indoor wireless optics communications solutions emerged as a relevant alternative.

3.2.1. *Outdoor wireless optical communication*

3.2.1.1. *Earth-satellite wireless optical communication*

The first known project was initiated in the 1970s [MAY 05]. It was called Spaceborne Flight Test System (Space Flight Test System) program and had the code number "405B Program". It was at that time "the study of a space laser communication" in the Weapons System (WS) projects, funded entirely by the U.S. Air Force and the Pentagon. This project was conducted with the participation of NASA, DARPA, Goddard Space Flight Center, and Jet Propulsion Laboratory.

During the same period and always with the financing of the U.S. military, telecommunications links between aircraft have been tested using high-power lasers. Then, logically, these experiments were extended in the 2000s with programs such as Oracle or Orcle. These programs proposed wireless optical communications prototypes from ground–ground, ground–plane, plane–airplane, satellite, aircraft, and satellite–submarine or high-altitude platform communication [GIG 02, HAP 11, HEN 05, HOR 04] or drones [LOH 11].

Finally, around 2005, studies were conducted on communications between Earth and modules on Mars (NASA's Space Exploration Initiative – SEI) with the

Lewis Research Center Laboratory. It was to use lasers in order to achieve data rates of 100–1,000 Mbps. The findings of preliminary studies showed the availability of a complete system for 2020 [KWO 92].

3.2.1.2. *Intersatellite wireless optical communication*

In the 1990s, civil studies were initiated for intersatellite communications, and in November 2001, the first civilian application of laser high-speed communication was implemented. A 50 Mbps link was established between the French satellite SPOT-4 and the European satellite Artemis, separated from the other by tens of thousands of miles. Figure 3.18 represents a wireless optics communication between SPOT-4 satellite and Artemis satellite, by using the Silex laser system.

This European success is the result of a project called Semiconductor Intersatellite Link Experiment (SILEX) initiated in the early 1990s. The partners were Matra Marconi Space, Astrium, the European Space Agency (ESA), and the Centre National d'Études Spatiales (CNES).

Since 2008, the German Space Agency (DLR) operates an intersatellite link between the satellite TerraSAR-X and NFIRE. This connection is based on a second generation of laser communications technology. This solution will also be used for the new European satellite data relay (European Data Relay Satellite – EDRS) in 2013. The low power consumption and compact size have favored this approach over the radio.

Figure 3.18. *Communication between SPOT-4 and Artemis with Silex laser system
(Source: ESA)*

3.2.1.3. *Free-space optic*

In France, the first trial of free-space laser communication was made in the late 1960s to Lannion between a building at the Centre National d'Études des Télécommunications (CNET) and trailer laboratory [TRE 67]. The device used a helium–neon laser of wavelength 632.8 nm at the emission side and a photomultiplier at the reception. Another study looked at a carbon dioxide laser wavelength of 10.6 μ. Video transmission tests were carried out over a distance of 1.2–19 km.

Other products, commercially mature, appeared in the world in the mid-1980s. But despite advances in technology, transmitters and sensors, transmission quality and availability of communication links did still not meet the expectations of a telecommunications operator.

But in the early 2000s, a new wave of products was proposed, mainly from European and American origin. These devices were tested by Centre Commun d'Études de Télévision et Télécommunications (CCETT) from Rennes (France) in order to determine what market segments the digital communications technology could propose.

The telecommunications market has become competitive; these technical solutions have been proposed as an extension to optical fiber or private Ethernet intersite communications.

The framework for using these wireless optical systems has been developed with the International Telecommunication Union (ITU) [ITU 11] and more specifically the ITU-R, CE3 [ITU 03, ITU 04, ITU 05b, ITU 07a, ITU 07b], CE5 [ITU 05a, ITU 08], and CE9.

3.2.2. *Indoor wireless optical communication*

The development of indoor wireless optics communication has existed since the early 1980s with a different approach to technical solutions according to the objectives, needs, and technological advances.

Table 3.3 outlines a non-exhaustive history of optical communications mentioning the company or the laboratory, the proposed data rate, the type of propagation, the modulation, and the used wavelength.

Wireless optical communications in limited space found many markets in domestic, transportation, and the office domain. Before dwelling on four typical examples, we can mention some everyday uses: wireless optical mouse and wireless keyboard, video game controllers with position sensing (indoor positioning – IdP), the wireless door key for access to homes or vehicles, the bar code reader and the electronic tag in the supermarket, and the wireless stereo headphones. There are also more innovative applications such as wireless optical communications between integrated circuits on a printed circuit board and the wireless optical network sensors in a plane [DAV 09].

Date	Organization	Data rate	Type	λ (nm)
1979	IBM Zurich	125 Kbps	Diffuse	IR 950
1985	HP Labs	1 Mbps	LOS	IR 660/880
1986	Motorola	50 Kbps	LOS	IR 950
1987	Bell Labs	45 Mbps	LOS	IR 950
1988	Matsushita	19 Kbps	Diffuse	IR 880
1993	BT Labs	155Mbps	LOS	IR 850
1997	IrDA – IBM	16 Mbps	LOS	IR 850
1998	JVC	10 Mbps	WLOS	IR 850
1999	Schneider	115 Kbps	LOS	VLC
2000	Keio University	1 Mbps	LOS	VLC
2001	JVC –VipsLan100	100 Mbps	WLOS	IR 850
2002	Schneider	4 Mbps	LOS	VLC
2003	IrDA	100 Mbps	LOS	IR 850
2005	JVC – Luciole	1.5 Gbps	LOS	IR 850
2006	Keio University	100 Mbps	LOS	VLC

Table 3.3. *Historic list of wireless optical communications*

3.2.2.1. *The remote controller*

By definition, a remote control is an electronic device for PtP unidirectional used to modify the remote operation of equipment, such as a TV channel and the volume. The first public television remote control appeared in 1955 (Zenith with Flash-Matic). It used visible light and four-cell reception (on/off, volume and the channels selection). One year later, wireless radio technology, with a greater coverage, replaced the light solution. But radio waves pass through obstacles such as walls, resulting in potentially changing the TV program in the neighborhood. The infrared solution was then finally retained.

Since the early 1980s, the vast majority of remote controls use infrared technology, and dispatch of orders is done by transmitting a digital signal, whose frequency is modulated at a few tens of kilohertz. Owing to the proliferation of owner controllable devices by a remote controller, up to 10 per household, there is now universal remote controller that controls all or part of the buttons. They are configurable or appear on an LCD touch screen.

3.2.2.2. *The visible light communication*

Visible Light Communication Consortium (VLCC) is a Japanese association, which aims to search, develop, and propose standardization of communication systems, using a ubiquitous solution, LED in the visible spectral range. Communication is achieved by intensity modulated at high frequency, with a consideration for human security. The use of communication is available in personal lighting, offices, car lights or road infrastructure (Intelligent Transport System – ITS), electronic signs for advertising, etc. The advantage of this approach is to offer a unique solution for lighting and communication. From 2007, papers were proposed and new working groups from the IEEE 802.15 [WON 07] have been created (IEEE 802.15.7) to provide communications solutions in the visible area with PtP and point-to-multipoint (PmP) solution.

3.2.2.3. *The IrDA solutions*

Another use case that comes from Infrared Data Association (IrDA) offers a 1 Gbps PtP solution called Giga-Ethernet (previously known as EFIR) for a distance of 0.2 m [KDD 11]. To meet the demands of rapid data transfer, the members of this association are intending to provide such a system in 2012. Examples of use are a portable multidevice interface and fast music or film download.

As part of a collaborative project called Regional Techimages, Orange Labs in Rennes, with regional research partners, have achieved a PtP at 1.25 Gbps over a distance of 0.5 m [BOU 09]. As part of a European collaborative project called Omega [OME 11a], the 1.25 Gbps data rate was achieved over a distance of 3 m.

3.2.2.4. *The indoor wireless optical network (WON)*

The VISPLAN product was the first system providing a wireless optical communication network, and it was only sold in Japan by JVC. VIPSLAN is typically a WLAN or a WDAN in direct competition with WiFi solutions. But the first version of WiFi pushed this device into the background. Based on the work of the Infrared Communication Systems Association (ICSA) [ICS 11] and the recommendations of the IR PHY IEEE 802.11 [IEE 11b], it offered a rate of 10 Mbps, then 100 Mbps.

The device (Figure 3.19) consisted of two elements:

– the base station (or COIL – Figure 3.19(a)) providing a data rate of 100 Mbps Ethernet with a range of 5 m (coverage of about 25 m^2) in WLOS propagation;

– modules (or MOIL – Figure 3.19(b)) with the same characteristics of speed and range mentioned above, but constituted of an automatic pointing device.

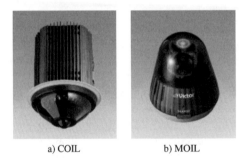

a) COIL   b) MOIL

Figure 3.19. *Visplan (Source: JVC)*

So far, in this last area, the home or business wireless network WDAN and WLAN, commercial success was not to progress and several reasons can be advanced:

– A non-economically viable offer for a PtM optical link budget in a room.

– A non-available or inadequate network m layer (MAC layer).

– A broadband access point (xDSL or FTTH) or a connectivity insuring an inter-room link and using the power line communication (PLC) technique or fiber optic was not available.

– The concepts of energy savings or safety were less important.

However, over the past decade, the increase in throughput was highly significant and did not seem to have reached its asymptote. WiFi (IEEE 802.11) is a perfect example.

In wireless communication, the growing need for throughput can make indoor wireless optics technology a good alternative or complementary solution to the radio systems because of some interesting features:

– A high throughput (>1 Gbps) can be achieved, like radio system (e.g. system at 60 GHz).

– A spectral availability of more than 700,000 GHz, unregulated or taxed.

– The optical transmission is limited within a room, so they are naturally more secure.

– It is possible to reuse the same optical wavelength in the next room or the neighboring apartment with the same level of security.

– The installation of equipment is more intuitive (optical propagation).

– There is no suffered or performed interference with radio systems.

– The safety aspect or immunity refers to recognized international standards (IEC or FCC 60825).

3.2.3. *The institutional and technical ecosystem*

Communication solutions in wireless optical communications in limited space can be divided into several areas, shown in Figure 3.20.

– VLC: The application VLC was first devoted to PtP, short-range, and low data rate solutions by the VLCC association. But due to the work done within the IEEE 802.15.7 Working Group, VLC specification became wider.

– IEEE 802.15.7: IEEE 802.15 committee focuses on the development of PtP and PmP standards of WLAN in the visible area. In this context, a task group (TG) dedicated to the VLC was established in January 2009. Called 802.15.7 and chaired by Samsung, this working group has proposed a standard in 2010 whose main specifications are:

- the PtP and PmP solutions with a star or an *ad hoc* network configuration;

- two physical layers (PHY layer): Type 1 low data rate (10–100 Kbps) and Type 2 high speed (3.2–96 Mbps);

- the low data rate applications are mainly for road signs information (ITS), the dissemination of information in public or domestic places, indoor geolocation and advertising messages;

- the applications for high speed concern diffusion in public space or for household purposes (music, video, etc.) or the fast downloading for mobile devices (PDA, phone, etc.).

– ECMA: The VLC was also discussed at the ECMA in 2009. To this end, a white paper presented by the Lumilink company was proposed to the Technical Committee 47 (TC47) and a presentation to Samsung to promote the potential of visible light LED for near field communication.

– IrDA: Established in 1993, IrDA is an association that works in the infrared. In 1997, the association suggested a recommendation for a PtP economic digital infrared module (Infrared Communication – IRC). The IrDA device was present on many portable devices such as phones and laptops and also on devices such as printers and video cameras. Several standards have been developed and these provide the increased flow (data rate). The final specification, completed in 2010, offers an Ethernet communication up to 1 Gbps. The goal is to offer this standard on the same equipment with various applications: multi-interface mobile devices, music or video downloading from a Kiosk or at the rental store, promotion or advertising message, etc. Members are Casio, Finisar, Fuji, KDDI, Mitsubishi, NEC, NTT, Panasonic, Sony Ericsson, etc.

– OWMAC: The Optical Wireless Media Access Control (OWMAC) protocol specification is clearly defined for PmP broadband in a home or business network [OME 09]. This specification is wavelength independent and can more integrate the IEEE 802.15.7 and IrDA applications, and home automation applications.

If we have to make a comparison with another technological solution, it is always a very delicate operation as it should, initially, harmonize the above mentioned characteristics definitions.

In addition, it is preferable and realistic to better define the different radio and wireless optical technologies in a complementary than a competitive aspect.

However, a qualitative comparison (Table 3.4) is available from a radio technology that has similar characteristics (systems in 60 GHz) in order to obtain a more general view on the advantages and disadvantages of these two technologies.

Indoor wireless optics network is much less mature than radio systems, but it can be seen as having some equivalences, although it is difficult to make a comparison to "equivalent basis".

Other parameters may be considered, such as energy use and safety. We now address the first aspect of the wireless optical communication, the transmission channel modeling.

Features	Radio 60 GHz	Wireless Optic
Spectrum availability	Reduced	Abundant
Spectrum regulation	Restricted	Free
Spectrum fee	Important to free	Free
Multipath fading	Very important	None
Data security	Encryption	Intrinsically secure by the walls
Intersymbol interference	Low	Potentially significant at high speed
Created or suffered electromagnetic interference	Possible if similar frequency or harmonic	None
Dominant noise	Other users	Artificial light and daylight
Human safety	Epidemiological study ongoing	Safety (Class 1) Internationally accepted standards

Table 3.4. *Comparison between 60 GHz radio and wireless optical technologies*

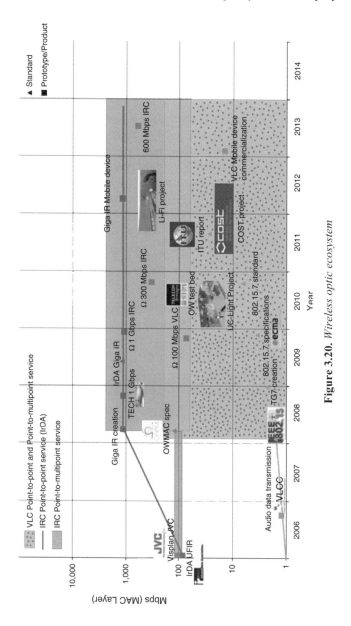

Figure 3.20. *Wireless optic ecosystem*

Chapter 4

Propagation Model

"Definition of free space propagation: In a homogeneous space comparable to the vacuum, the radiated energy of an isotropic source propagates at light speed and uniformly spread over the surface of a sphere whose radius increases with time."
Fundamentals of Radio Propagation,
Jacques Deygout, Eyrolles, Paris, 1994

4.1. Introduction

The aim of this chapter is to give a description of a baseband wireless optical link. First, we compare the modeling of the system to a radio system. Then, we describe the disruptive noises and define the electrical signal-to-noise ratio (SNR). Finally, an analysis of the diffuse channel is performed and different calculation models are presented.

4.2. Baseband equivalent model

In this section, the radio propagation principles are first remembered and compared to those of free-space optical ones. Finally, a baseband free-space optical link model is presented.

4.2.1. *Radio propagation model*

We can define a real signal $f(t)$, located in a narrow frequency band around a carrier frequency f_0. When the frequency band occupied by $f(t)$ is small compared to f_0, the signal can be called narrowband signal or passband signal. It can be written in terms of its complex envelope in the following form:

$$f(t) = \text{Re}\left\{a_f(t)e^{2j\pi f_0 t}\right\}$$

with: $a_f(t) = p(t) + jq(t) = r(t)e^{j\phi(t)}$

Equation 4.1. *Narrowband signal (1st representation)*

where f_0 is the carrier frequency, $a_f(t)$ is the complex envelope of the signal, and $p(t)$ and $q(t)$ are quantities, respectively, called phase and quadrature components of $f(t)$. They are real values and low-pass-type signals, generated in order to carry digital information by choosing a given type of modulation (OOK, PSK, QAM, etc.).

The development of equation [4.1] can write the narrowband signal $f(t)$ in the following two equivalent forms:

$$f(t) = p(t)\cos(2\pi f_0 t) - q(t)\sin(2\pi f_0 t)$$

$$= r(t)\cos(2\pi f_0 t + \phi(t))$$

Equation 4.2. *Narrowband signal (2nd representation)*

This approach is interesting because it can represent any narrowband or passband signal from its phase and quadrature components [PRO 00].

The $\cos(2\pi f_0 t)$ and $\sin(2\pi f_0 t)$ signals are produced on the emission side by a local oscillator at the frequency f_0 (Figure 4.1). At the reception, a similar local oscillator allows us to find the phase and quadrature components from the signal $f(t)$. It must be in that case phase and frequency synchronized with the received signal carrier.

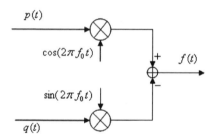

Figure 4.1. *General form of narrowband signal*

The transmission channel is defined as the physical medium of propagation. For wireless links, it is, for example, the room itself. By neglecting the channel time variations and ignoring the noise, the received signal is then expressed as:

$$v(t) = \mathrm{Re}\left\{ \rho(t) e^{2j\pi f_0 t} \right\}$$

where

$$\rho(t) = \sum_{k=0}^{N-1} c_k a_f(t - t_k) e^{j\theta_k}$$

Equation 4.3. *Received useful signal*

In this equation (equation [4.3]), c_k, t_k and θ_k are, respectively, the amplitude, time delay, and phase change for the order k component of the path (kth path), and N is the total number of received paths.

The phase θ_k depends on the length of the covered path. For each distance equal to the signal wavelength, the phase change modulo is 2π so that the phase of the received signal can vary significantly and the resulting sum may cause signal loss. This is important in radio propagation.

For example, at the radio frequency of 1 GHz, signal fluctuations are separated in distance by 30 cm. The received signal on a small antenna spatially varies due to multipath and that antenna can be located in an area that causes a fading of the received signal. Thus, for each position of the equipment in a room, it is possible to model the channel as a linear filter.

4.2.2. *Model of free-space optical propagation*

In the field of free-space optic, it is expensive and difficult to control the phase of an optical carrier; moreover, in case of reflection, there is a partial or a total change of this phase. Furthermore, in reception, it is also difficult to estimate the phase in free-space optic. This is why most current equipments use techniques of light intensity modulation (IM) such as the On/Off Keying (OOK) and the pulse position modulation (PPM). The binary data are transmitted by the instantaneous optical power of the emission device.

This optical emission device is generally a light-emitting diode (LED) or a laser that converts an electrical current proportionally to an optical power.

Similarly, the reception is based on direct detection (DD) in which the photodetector produces a current proportional to the instantaneous received optical power. This type of transmission is commonly known as IM/DD.

Multiple reflections can lead to spatial variations in the amplitude and phase of the received optical signal. In the field of radio, multipath fading is known because the size of the detector is smaller than the wavelength. Although, in the field of free-space optic, multipath fading is present but the size of the detector being very large compared to the wavelength, we get signal integration.

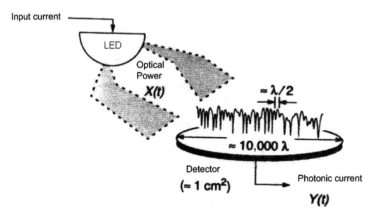

Figure 4.2. *Photo electrical detection and wavelength*

Figure 4.2 [KAH 97] describes an infrared channel using IM/DD. It consists of an infrared transmitter and a significant photoreceptor area. The dimensions of a typical photodetector with an area of 1 mm^2 are several orders of magnitude compared to the optical wavelength of the signal, which is of the order of 1 μm. The output signal of the photocell ($Y(t)$), which acts as an integrator, is an instantaneous spatial average of the received optical signal, including all impairments values.

Nevertheless, in time domain, since the emitted optical power is spread over several optical paths of different lengths, wireless optical communication can suffer from multipath distortion or intersymbol interference (ISI). From a certain data rate, from 20 Mbps to over 1 Gbps, function of the volume of the room, the location and orientation of the transmitter and receiver, this characteristic may be detrimental in the case of a diffuse communication.

For systems using light IM at the emission with DD at the reception (IM/DD), the propagation medium can be replaced by an equivalent baseband model [HAS 94, BAR 94, KAH 97]. The source is assumed to be linear and we can write:

$$y(t) = R.x(t) \otimes h(t) + n(t) = z(t) + n(t)$$

Equation 4.4. *Equivalent baseband model*

where

– $y(t)$ is the instantaneous current delivered by the photodetector (A).

– R is the photodetector responsivity or the spectral sensitivity (A/W).

– $x(t)$ is the instantaneous optical power of the emitter (W). This optical power is determined with respect to the used modulation.

– $h(t)$ is the channel impulse response invoking the N paths.

– $\otimes$ is the convolution operator.

– $n(t)$ is the additive white noise. It can be regarded as a Gaussian noise independent of $x(t)$.

From the model described above (equation [4.4]), all the channel coefficients represent the attenuations and their values are real and positive. It is shown that although the information is carried by an optical carrier, the whole system can be expressed as an equivalent baseband model (Figure 4.3).

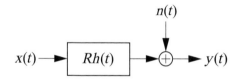

Figure 4.3. *Equivalent baseband model*

In a confined space configuration, the channel is assumed static. This approach is generally appropriate to model this channel, since we have a change only when the transmitter or the receiver is moving, even when objects or people in the room move or are moved [KAH 97].

So the input–output relation of the channel is a convolution, with the particularity that the function $x(t)$ represents an instantaneous power, so it is not negative ($x(t) \geq 0$).

From this property, the temporal average of the signal is a positive quantity; it takes the form of the transmitted average optical power:

$$P_t = \lim_{T \to \infty} \frac{1}{2T} \int_{-T}^{+T} x(t)dt$$

Equation 4.5. *Transmitted average optical power*

so that the received average optical power is proportional to the temporal integral of $x(t)$ and is expressed as:

$$P_r = \lim_{T \to \infty} \left\{ \frac{1}{2T} \int_{-T}^{+T} \frac{1}{R} z(t)dt \right\} = \lim_{T \to \infty} \left\{ \frac{1}{2T} \int_{-T}^{+T} \left(x(t) \otimes h(t) \right)dt \right\} = H(0)Pt$$

Equation 4.6. *Received average optical power*

where $H(0)$ represents the direct current (DC) channel gain (frequency equal to zero) or the useful average optical power attenuation. Its expression is:

$$H(0) = H(f)_{f=0} = \int_{-\infty}^{+\infty} h(t)dt$$

Equation 4.7. *Direct current channel gain*

It should be noted that from equation [4.7], the received average optical power is proportional to the average value of $z(t)$, unlike the electricity domain where the average power of the useful signal to the output of a photodiode is [WOL 03 and LAL 07]:

$$P_{elec} = E\left[\left|z(t)^2\right|\right] = E\left[z(t)^2\right] = R^2 \|h\|^2 \left(\sigma_x^2\right) + R^2 H(0)^2 P_t^2$$

Equation 4.8. *Average electrical power of the useful signal*

The average of the signal $x(t)$ is the average of the transmitted optical power P_t and the value (σ_x^2) is its variance.

Unlike most radio systems, the average of the useful signal is not equal to zero and it adds to the total electric transmitted power.

The last parameter is the noise $n(t)$ that can be divided into three main sources:

– The thermal noise due to thermal effects in the receiver.

– The periodic noise, or optical origin electrical disturbance, is mainly the result of the variation of light due to the design method of the lamp using an electronic switching power supply. This produces a specific periodic signal with, most of the time, a fundamental frequency of 44 kHz and significant harmonics up to several hundred kilohertz. All ambient artificial light sources are modulated at low frequency (mains frequency: 50 or 60 Hz) or high frequency (in the case of fluorescent lamps).

– The shot noise or optical origin disturbances that are the result of photon arrivals from the useful signal and ambient light sources such as the lamp lights and the Sun.

In the receiving wireless optical device, the shot noise usually dominates when the device is well made [KAH 97, TAN 02, ALQ 03a]. It is defined by:

$$\sigma_t^2 \cong \sigma_{Noise}^2 = N_0^{ss} \times B$$

Equation 4.9. *Noise power expression*

The B term is the modulation bandwidth; it can be defined from a percentage of power of the first zero frequency or from the Nyquist criterion. In the case of a binary modulation such as OOK modulation, the bandwidth B is equal to the bit rate.

The term N_0^{ss} is the single-sided power spectral density. Theoretically, the shot noise comes only from the useful incident beam and from ambient light entering the detector active area (photodiode). However, this noise from ambient light is often very high compared to the useful signal, despite the use of a suitable optical filter in the receiver. Therefore, it is possible to neglect the shot noise generated by a useful signal.

In addition, since most disturbing ambient light has a mainly continuous component, the integration of a dynamic shift or a suitable filter in the receiving device offers a first simple and inexpensive solution.

Finally, the induced shot noise from the ambient light can be modeled as a Gaussian noise [KAH 97, ALQ 03a]:

$$N_0^{SS} = 2q(I_{inc} + I_d) = 2q(I_u + I_{bg} + I_d) \cong 2qI_{bg} = 2qRP_{bg}$$

Equation 4.10. *Spectral density of dominant noise*

where

- $q = 1.6 \times 10^{-19}$ is the electron charge.

- I_{inc} is the current produced by the incident optical power.

- I_d is the dark current of the photoreceptor.

- I_u is the useful signal current.

- I_{bg} is the background light current.

- P_{bg} is the optical power of the ambient light.

Four main types of ambient light can interfere with wireless optical communication in domestic and professional environments:

- incandescent lamps;

- fluorescent lamps;

- devices with LED; and

- the Sun.

Several experiments were carried out. A synthetic example is presented (Figure 4.4), a representation of the disruptive effect in function of the wavelength [BOU 07], at 1 m distance from the source (except the Sun).

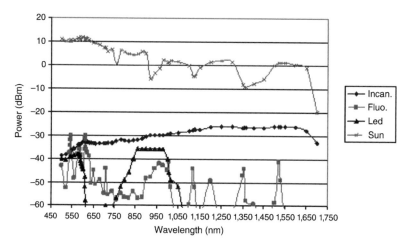

Figure 4.4. *Example of optical disruptors*

It appears that the Sun in line of sight is the most important disruptor, but its influence may be significant even for non-line of sight. Then, in decreasing order are the incandescent lamp (which are set to disappear in the near future), fluorescent lamps, and LEDs.

As part of the disruptor management, an indication of the levels can capture the preferred spectral ranges for wireless optical communication device.

4.2.3. *The signal-to-noise ratio*

To study the system performances and compare them to other systems based on different technologies of digital communications, it is possible to define a parameter that is the SNR, ratio of the useful electric power to the noise power:

$$SNR = \frac{P_{Elec}}{P_{Noise}} = \frac{E\left[\left(x(t)\otimes h(t)\right)^2\right]}{\sigma_t^2} = \frac{E\left[z(t)^2\right]}{\sigma_t^2}$$

$$SNR = \frac{R^2\|h^2\|(\sigma_x^2)+R^2P_r^2}{\sigma_t^2} = \frac{R^2\|h^2\|(\sigma_x^2)+R^2H(0)^2P_t^2}{\sigma_t^2}$$

Equation 4.11. *Electric SNR*

Since the received useful signal current is directly proportional to the received optical power, the electric SNR commonly used in many studies in the wireless optical field [KAH 97, TAN 02, CAR 02a] only takes into account the square of the average received signal as an input signal and not the average of the square of the signal.

This expression considers only the variable part of the modulated signal and removes the DC component related to the fact that the received signal does not have a null average.

$$SNR_{Optic} = \frac{E[z(t)^2]}{\sigma_t^2} = \frac{R^2 P_r^2}{\sigma_t^2} = \frac{R^2 H(0)^2 P_t^2}{\sigma_t^2}$$

$$SNR_{Optic} \cong \frac{R^2 H(0)^2 P_t^2}{2qRBP_b}$$

Equation 4.12. *Electrical SNR in wireless optical communication*

In the case of line of sight propagation, theoretically, the only noise taken into account is the shot noise from ambient light. It is equivalent to an additive white Gaussian noise.

For example, a typical value of the spectral radiance of ambient light for wavelengths approximately 850 nm, during a sunny day and inside a house, is 0.04 μW/(mm^2.sr.nm) [WOL 09]. The noise value will therefore depend strongly on the effective area of the optical receiver, the optical filtering, and the field of view (FOV) or solid angle.

In fact, it should also take into account the unwanted effects of electronic circuits, especially in the choice and implementation of amplifiers, preamplifiers, and (or) photodiode(s).

Nevertheless, in wireless optical transmission scheme, there is an important difference with radio systems. The electrical SNR depends on the square of the average transmitted power Pt and not only on Pt because it is evaluated after the quadratic detection (optical/electrical conversion). So when it is attenuated by 3 dB, the SNR is reduced by 6 dB, and conversely, when the optical power is increased by 3 dB, the SNR reaches 6 dB gain. This parameter is important in finding solution to increase the gain in a link budget.

4.3. Diffuse propagation link budget in a confined environment

In this section, we present ISI and optical reflection models on materials before discussing the various models proposed in diffuse propagation.

4.3.1. *Intersymbol interference*

In diffuse propagation, the receiver receives not only the optical beam of the direct path, but also the reflected path beams on different surfaces of the room. For example, in OOK modulation case, if the frequency of the symbols becomes very important, we will have an overlap of values that will produce an interference phenomenon; this phenomenon is called ISI.

For reliable communication (i.e. with an ISI close to zero), it is necessary that the entire system takes into account the impulse response requirements.

An example is given in Figure 4.5 with three different paths:

– the direct path (factor $h = 0$), "direct";

– the one reflection path (factor $h = 1$), "1 reflection";

– the two reflections paths (factor $h = 2$), "2 reflections".

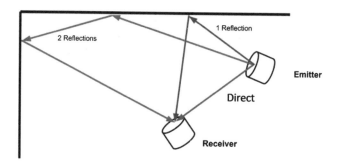

Figure 4.5. *Multipath*

In this example and when the symbols' frequency becomes very important, we get an overlap of values as shown in Figure 4.6. The impulse response of the first symbol (solid line) includes the direct path (H0), the path with one reflection (H1), and a third path with two reflections (H2).

The impulse responses of higher order ($n>3$) are generally considered as negligible. The impulse response of the first symbol can disrupt the second (dotted line) and the third (line and two points) symbols because despite the fact that it is a lower amplitude signal, its value may be sufficient to degrade the quality of transmission.

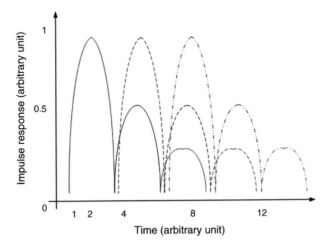

Figure 4.6. *Example of overlapping symbols*

Figure 4.7 gives us an example of the impulse response of the incident path and the different reflections for two different rooms configurations (2 m × 2 m – line and 4 m × 4 m – dotted line); the respective distances between transmitter and receiver are the same.

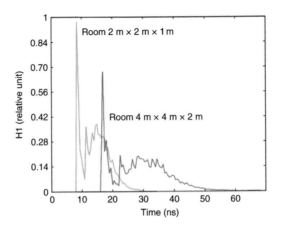

Figure 4.7. *Example of impulse response in confined environment*

One of the first impulse response models was proposed by Gfeller [GFE 79]:

$$h(t) = \frac{2t_0^2}{t^3 \sin(FOV)} \quad \text{if } t_0 < t < \frac{t_0}{\cos(FOV)}$$

otherwise $h(t) = 0$

where T_0 is the minimum delay between the source and the receiver (the shortest path delay).

Equation 4.13. *Impulse response model*

Figure 4.8 shows the shape of the impulse response $h(t)$ for two values of the FOV angle (1 = $\pi/6$ or 30° (triangle) and 2 = $\pi/4$ or 45° (square), where $t_0 = 2.10^{-9}$ s.

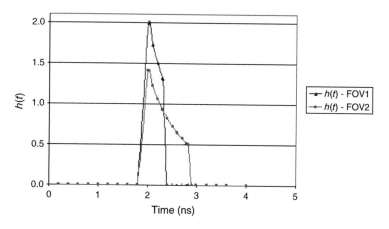

Figure 4.8. *Example of impulse response h(t) for two FOV values*

The temporal width of the impulse response is proportional to FOV angle. This behavior is logical because when this angle (which characterizes the receiver) is decreasing, there are fewer reflected rays on the walls of the room or on the furniture that can be received (which arrive at the receiver under an angle with its axis below the receiving system angle (FOV)). When FOV increases, wide line of sight and diffuse propagation case, the reflected paths number and the reflected paths power arriving at the receiver increase proportionally. The response delay is then greater and it is possible to obtain the symbol $T+1$ in correspondence with reflections of the symbol T. In this case, then there is occurrence of interference between symbols; the result is a performance degradation.

This degradation appears to occur at a data rate between 50 Mbps [OBR 04] and 260 Mbps [GFE 79] depending on the room configuration and the respective positions of the transmitter and the receiver. However, solutions, such as dynamic equalization [KAH 95] and OFDM [ELT 09], from the radio channel or the optical fiber could help overcome this problem.

The next step is the reflection modeling of an optical beam and the presentation of different models to characterize the impulse response including the most accurate and the most complete elements mentioned above. Several models exist and they are mentioned in the following sections.

4.3.2. Reflection models

The phenomenon of reflection occurs when a wave comes against a surface with large dimensions compared to the wavelength (floor, wall, ceiling, furniture, etc.).

The reflection characteristics of any surface depend on several factors:

– the material surface (smooth or rough);

– the wavelength of the incident radiation; and

– the angle of incidence.

4.3.2.1. Specular reflection

The roughness of the surface of a structure compared to the wavelength of the incident signal constitutes an important parameter of the shape of the reflection pattern.

A smooth surface reflects the incident radiation in one direction like a mirror and Descartes' law is applied, the reflection is called specular reflection.

But, unlike a radio channel for which the reflections on the surfaces are mostly specular type, in optics field, the dominant reflections are diffuse type.

4.3.2.2. Diffuse reflection

In the case of a rough surface, the incident radiation will be reflected in all directions. A surface is considered as rough, according to the Rayleigh criterion, if the following relationship is satisfied:

$$\varsigma > \frac{\lambda}{8\sin\theta_i}$$

Equation 4.14. *Rugosity criteria*

where

   – ς: maximum height of surface irregularities;

   – λ: wavelength of incident radiation;

   – θ_i: angle of incidence.

For optical radiation of 1,550, 850, and 550 nm wavelengths, assuming normal incidence, a surface is rough if the maximum height of the surface irregularities ς is, respectively, greater than 0.19, 0.11, and 0.07 μ.

These values mean that most of the surfaces found inside buildings may be considered as rough for optical radiation. In this case, the reflection pattern presents a high diffuse component, the reflected wave is scattered in multiple directions, and this reflection is called diffuse reflection.

To integrate this parameter, two models are commonly used to represent the reflection of optical radiation: Lambert's and Phong's models.

4.3.2.3. Lambert's model

Most of the surfaces are very irregular and reflect the optical radiation in all directions, independently of the incident radiation. Such surfaces are known as diffuse and can be approximated using Lambert's model. Figure 4.9 represents an experimental reflection pattern for an incident angle of 45° of a raw cement surface (before and after white painting) and their approximations according to Lambert's model).

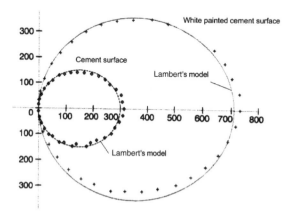

Figure 4.9. *Experimental reflection patterns of a rough cement surface (before and after white painting)*

This model is very simple and easy to implement using computational software, and the relation is:

$$R(\theta_0) = \rho R_i \frac{1}{\pi} \cos(\theta_0)$$

Equation 4.15. *Lambert's relation*

where

    – ρ: surface reflection coefficient;

    – R_i: incident optical power;

    – θ_0: observation angle.

Table 4.1 provides example of reflection coefficient values of an infrared beam for various material surfaces [YAN 00].

Material	Reflection coefficient
Painted wall	0.184
Painted paper	0.184
Wooden floor	0.128
Brown wood shelf	0.0884
Clear glass	0.0625
White ceramic	0.0517
Plastic	0.1018

Table 4.1. *Reflection coefficient ρ*

For example, if we consider a painted wall (ρ = 0.184), a receiver located 20° from the normal to the wall, and an incident power equal to 10 mW, the reflected power value ($R(\theta)$) will be:

$$R(20°) = 0.184 \times 10 \frac{1}{\pi} \cos(20°) = 0.55 \text{ mW}$$

Equation 4.16. *Example of Lambert's reflection*

4.3.2.4. *Phong's model*

The reflection pattern of several rough surfaces is well represented by Lambert's model, but it is less valid under certain materials incidences on which specular reflection has a significant component. Phong's model considers the reflection pattern as the sum of two components: the diffuse and specular component. The percentage of each component mainly depends on the surface characteristics and is a parameter of the model. The diffuse component is modeled by Lambert's model and is incorporated into Phong's model. The specular component is modeled by a function that now depends on the angle of incidence θ_i and the observation angle (or the reflection angle) θ_0. The Phong's model is described by the relation:

$$R(\theta_i,\theta_0) = \rho \frac{R_i}{\pi}\left[r_d \cos(\theta_0) + (1 - r_d)\cos^m(\theta_0 - \theta_i)\right]$$

Equation 4.17. *Phong's relation*

where

− ρ: surface reflection coefficient;

− R_i: incident optical power;

− r_d: percentage of incident signal that is diffusely reflected (it is a value ranging between 0 and 1);

− m: parameter that controls the directivity of the specular component of the reflection;

− θ_i: incidence angle;

− θ_0: observation angle.

NOTE:− Lambert's model appears as a special case of the Phong's model taking $r_d = 1$.

Many reflection pattern measurements are found in the literature. In Figure 4.10, which represents the reflection patterns of a varnished wood surface for angles of incidence equal to 0° and 45° and their approximation by Lambert's and Phong's methods, authors [LOM 98] present reflection graphs of the different materials in polar diagram form.

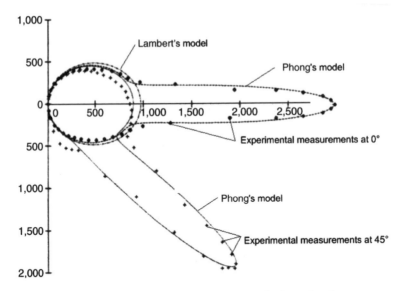

Figure 4.10. *Reflection patterns for a varnished wood surface*

Other reflection patterns exist in the literature and result from experimental measurements carried out by many authors [NIC 77, PHO 75, YAN 00]. The Phong's model equations and reflection coefficients are given in Table 4.2 for different types of materials: painted paper, wood, etc.

Materials	Phong's equation
Painted wall	$P(\theta_0) = \cos(\theta_0)$
Paint paper	$P(\theta_0) = \cos(\theta_0)$
Glass	$P(\theta_i, \theta_0) = 0.001\cos(\theta_0) + (1-0.001)\cos^{13}(\theta_0 - \theta_i)$
White ceramic	$P(\theta_i, \theta_0) = 0.06\cos(\theta_0) + (1-0.06)\cos^{1}(\theta_0 - \theta_i)$
Varnished wood	$P(\theta_i, \theta_0) = 0.3\cos(\theta_0) + (1-0.3)\cos^{97}(\theta_0 - \theta_i)$
Formica	$P(\theta_i, \theta_0) = 0.14\cos(\theta_0) + (1-0.14)\cos^{112}(\theta_0 - \theta_i)$
Plastic	$P(\theta_i, \theta_0) = 0.55\cos(\theta_0) + (1-0.55)\cos^{3}(\theta_0 - \theta_i)$

Table 4.2. *Reflection coefficient of different materials*

4.3.3. *Modeling*

As in all telecommunications systems, an analysis of the propagation channel is essential. As part of an optical communication in a confined space, the channel is characterized by the room and its constituent elements. Several methods are proposed to simulate and evaluate the behavior of the diffuse optical signal and, in particular, the impulse response.

To calculate the impulse response $h(t)$ of the channel, taking into account the diffuse component, it is necessary to resort to numerical simulations to take account of the light beams that reach the receiving device from multiple reflections on the different surfaces of the room and different furniture.

In 1979, Gfeller and Bapst [GFE 79] proposed a first single reflection propagation model. This model estimates the impairment errors introduced by using the reflection Lambert's model. The reflecting surface is divided into a large set of small elementary reflective areas called reflector elements. The spatial and temporal distributions of the transmitted signal are evaluated for each of these elements. Each of these elements is then regarded as a point source that emits the collected signal affected according to its reflection coefficient. The reflection pattern of each element is represented by Lambert's model described above. The received signal is the sum of the signals that arrive at the receiver after reflection on the different elements. Owing to the difference in propagation length of the various optical paths, the received signal is dispersed over time. In this model, only the effects of the reflected model on the path loss without considering the signal propagation delays are evaluated. The accuracy of the model increases with the reduction in the size of the reflective area of each element. The principle remains the same for the following models but they introduce in the simulation several consecutive reflections.

In 1993, Barry [BAR 93] proposed a recursive simulation model for multiple reflections. This model is limited to an empty rectangular room and is based on the reflection of Lambert's model. In 1997, Perez-Jimenez *et al.* presented a statistical model to estimate the impulse response [PER 97], while Lopez-Hernandez proposed DUSTIN algorithm [LOP 97]. In 1998, the Monte Carlo model, integrating the production of random radiation, was proposed for the simulation of a multipath wireless optical system in confined space [PER 98a, PER 98b]. Carruthers and Kannan have proposed a new iterative model based on an estimate of the impulse response [CAR 02b].

A comparison [SIV 03] of these techniques has shown that for a similar processing time, the Monte Carlo model seems more accurate than the DUSTIN algorithm. The Monte Carlo model is able to generate a simulation considering up to

40 reflections, which offers better accuracy. In these methods, the simulation is performed in three steps: (1) the ray generation, (2) the walls characterization, and (3) the photodiode response determination. The rays are randomly generated from the transmitting device and according to its transmission pattern. The impact points of each ray are then calculated. These impact points are either on the receiver effective area or on one wall of the room.

If the impact point is on a wall, then this point is considered as a new secondary light source and a new ray is generated based on the same method as before. If the point of impact reaches the receiver, then the incident power is saved as a vector with an appropriate position, which is calculated by estimating the total delay of the optical path between the transmitter and the receiver. In the modified Monte Carlo model [PER 98b], the contribution of the direct path between the source (wall or transmitter) and the receiver is calculated each time (Figure 4.11) and this greatly speeds up the processing time.

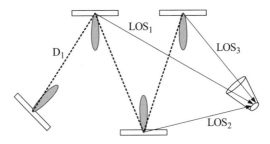

Figure 4.11. *Schematic view of the modified Monte Carlo model*

Nettle's model is slightly different from previous models; this model [NET 99] has its origin in the process of iterative calculation of a diffuse surface [GOR 84]. Radiosity is a lighting technique used in some three-dimensional (3D) computer models. This is called global illumination because the illumination of each elementary surface can be calculated separately from the others. The system modeling all the lighting or energy transfers can be globally solved. Radiosity uses the light radiative transfer physical formulas between the different elementary diffuse surfaces that compose a 3D scene.

For example, a simple scene can be modeled using a polygonal mesh, e.g. a cubic empty room divided into elementary surface elements. These basic elements used by the radiosity are patches (plane surface elements). Each of the faces of the cube and the room is a patch (Figure 4.12).

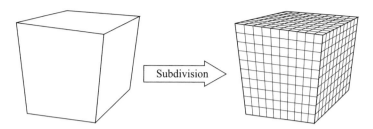

Figure 4.12. *Subdivision of a 3D cube into patches*

Each of these patches receives energy from the other sides, absorbs a certain amount (depending on the material properties of the patch), and returns the other part to the other patches. The energy transmitted from a patch A to a patch B is a function of the following parameters:

– the respective value of the normal to patches;

– a vector representing the direction of the center of the emitting patch toward the center of the receiving patch;

– the average distance between two patches;

– the respective surface of the patches;

– the transmitted optical power; and

– the level of visibility between two patches (obstacle).

Some patches are defined as emission optical source. For each transmitter, we determine the receiving patches.

The radiosity process is recursive and illuminated patches at the current iteration will be emission optical sources for the next iteration. Specifically, the radiosity emitted by a patch i, (Bi), is equal to the self-emitted energy (Ei) added to the addition of all radiosity received from other patches j (Bj) weighted by a reflection factor depending on the material (Ri). The energy received by the plan patch i from patch j is equal to the radiosity emitted by j multiplied by a form factor (Fi-j), depending on the relative orientation of i and j, their distance, and the presence of other objects (obstacles) between the two patches.

It should be noted that the influence of the iterative component of order 4 and higher is negligible, and some authors consider that the component of order 3 or order 2 can also be considered negligible.

This computing model allows us the treatment of a 3D scene including objects and application examples provided in Chapters 6 and 11. The process is as follows:

– After creating a two-dimensional (2D) room, it is associated with a 3D scene generated by what is called a 3D engine (Ogre open-source module, for example);

– While processing, room data are transferred to the Nettle propagation module that get back the tree and the structure of 3D model. This operation is achieved by converting data into 2D triangular elementary objects and identified by their position in three dimensions.

– While calculating, for each emission source and each iteration, the elementary surfaces can be lit according to the room geometry and its constituent determined elements.

The result is done by:

– the transfer of the 2D propagation calculation process results to the 3D engine;

– the generation of a new 3D image file including the line of sight and diffuse optical propagation results.

Chapter 5

Propagation in the Atmosphere

"Look at the light and consider its beauty.
Closes and opens the eyes quickly:
what you see is already passed
and what you saw is gone"
Leonardo da Vinci,
Codex Forster,
1493–1505

5.1. Introduction

Laser beams, used to operate free-space optical (FSO) links, involve the transmission of an optical signal (visible or infrared) in the atmosphere. They interact with various components (molecules, aerosols, etc.) of the propagation medium. This interaction is at the origin of many phenomena such as absorption, diffusion, refraction, and scintillation. The only limitation known is heavy fog and they cannot cover distances more than a few kilometers. They are therefore suitable for the construction of networks between nearby buildings. The laser beams generally used have low power; and the environmental impact is negligible.

One of the challenges to take up is a better understanding of atmospheric effects on propagation in this frequency spectrum, to better optimize wireless broadband communications systems, and to evaluate their performance. It is a prerequisite to test communication equipments.

Atmospheric effects on propagation, such as absorption and molecular and aerosol diffusion, scintillation due to the change in the air index under the effect of temperature variation, hydrometeors attenuation (rain, snow), and their different models (Kruse and Kim, Bataille, Al Naboulsi, Carbonneau, etc.) are presented and confronted against experimental results.

The runway visual range (RVR), a parameter for characterizing the transparency of the atmosphere, is defined and various measuring instruments such as transmissometer and scatterometer are described.

5.2. The atmosphere

In the context of FSO links, the propagation medium is the Earth's atmosphere. It may be regarded as a series of concentric gaseous layers around the Earth. From 0 to approximately 80–90 km in altitude, we have the homosphere and, extending beyond these altitudes, the heterosphere. When considering the temperature gradients as a function of the altitude, homosphere can be seen with three layers: the troposphere, the stratosphere, and the mesosphere.

In the case of FSO communications, the focus is particularly on the troposphere because most of the meteorological phenomena occur in this layer. The propagation of light in this layer is influenced by:

– the gaseous atmospheric mixing;

– the presence of small suspended particles of various sizes (0.1–100 μm), the aerosols;

– the hydrometeors (rain, snow, and hail);

– the lithometeors (dust, smoke, and sand);

– the changes in the gradient of air refractive index (propagation medium) which is at the origin of scintillation and turbulence.

5.2.1. *The atmospheric gaseous composition*

To characterize the properties of atmospheric transmission affecting optronic systems, the atmospheric gaseous components are classified into two categories: the fixed density and the variable density proportion components.

The fixed density proportion components (variation less than 1%) are the majority components. They have a quasi-uniform distribution for altitudes ranging between 15 km and 20 km. These include, for example, O_2, N_2, Ar, and CO_2. In the visible and infrared wavelengths range up to 15 μm, CO_2 gases give the only important absorption lines.

The variable density components are the minority and their concentration depends on the geographical location (latitude and altitude), the environment (continental or maritime), and the weather conditions.

Water vapor is the main component of the atmosphere. Its concentration depends on climatic and meteorological parameters. For example, in marine areas, its concentration can reach 2%, while above 20 km altitude its presence is negligible.

The water vapor concentration is determined from the atmospheric humidity and can be defined in three different ways:

– Absolute humidity (g/m^3): the mass of water vapor per unit air volume.

– Relative humidity (%): the ratio between the absolute humidity and the maximum quantity of water vapor that could be contained in the air at the same temperature and at the same pressure.

– Number of mm of precipitable water (w_0) per unit distance, usually per km.

Another major variable component is ozone (O_3) whose concentration also varies with altitude (maximum content at 25 km), latitude, and season. It presents an important absorption in the ultraviolet and infrared radiation.

5.2.2. *Aerosols*

Aerosols are extremely fine solid or liquid particles, suspended in the atmosphere and with a very low fall speed. Their size generally lies between 0.01 μm and 100 μm. Owing to the action of Earth's gravity, the biggest particles $(r > 0.2$ μm) are in the vicinity of the ground. Fog and mist are liquid aerosols, and salt crystals and sand grains are solid aerosols.

These aerosols may induce severe disturbances in the propagation of optical waves since their dimensions are close to the wavelength. This is not the case for the millimeter (EHF waves), decimeter (SHF waves), meter (UHF waves), and decameter (HF waves) radio waves.

5.3. The propagation of light in the atmosphere

The characteristic performance of FSO links for data transmission depends on the environment, the atmosphere in which they propagate. The atmosphere, due to its composition, interacts with the light beam (optical or infrared): absorption and molecular and aerosol scattering (fog, hydrometeors – rain, snow) and scintillation due to the variation of the air index as a result of temperature variations.

Atmospheric attenuation results from an additive effect of absorption and scattering of the infrared light by gas molecules and aerosols present in the atmosphere. It is described by Beer's law, giving the transmittance as a function of distance:

$$\tau(d) = \frac{P(d)}{P(0)} = e^{-\sigma d}$$

where

— $\tau(d)$ is the transmittance at a distance d from the emitter.

— $P(d)$ is the signal power at a distance d from the transmitter.

— $P(0)$ is the transmitted power.

— σ is the attenuation or the extinction coefficient per unit of length.

Equation 5.1. *Beer's law*

Attenuation is related to transmittance by the following expression:

$$Aff_{dB}(d) = 10\log_{10}\left(1/\tau(d)\right)$$

The extinction coefficient σ is the sum of four terms:

$$\sigma = \alpha_m + \alpha_n + \beta_m + \beta_n$$

where

— α_m is the molecular absorption coefficient (N_2, O_2, H_2, H_2O, CO_2, O_3, etc.), note that this refers to the structure and the composition of the atmosphere.

— α_n is the absorption coefficient of the aerosols (fine solid or liquid particles) present in the atmosphere (ice, dust, smoke, etc.).

— β_m is the Rayleigh scattering coefficient resulting from the interaction of light with particles of size smaller than the wavelength.

— β_n is the Mie scattering coefficient; it appears when the incident particles are of the same order of magnitude as the wavelength of the transmitted wave.

Equation 5.2. *Extinction coefficient*

Absorption dominates in the infrared whereas scattering dominates in the visible and ultraviolet regions.

5.3.1. *Molecular absorption*

Molecular absorption is resulting from an interaction between radiation, atoms, and molecules of the medium (N_2, O_2, H_2, H_2O, CO_2, O_3, A_r, etc.). It defines different transmission windows in visible and infrared frequencies domain (Figure 5.1).

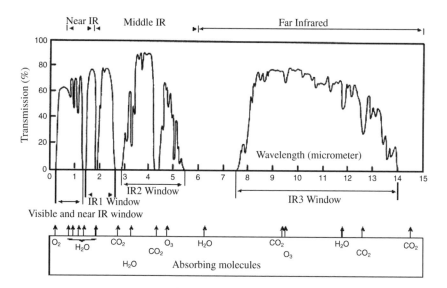

Figure 5.1. *Transmittance of the atmosphere due to molecular absorption*

5.3.2. *Molecular scattering*

It results from the interaction of light with atmospheric particles whose sizes are smaller than the wavelength. An approximate value of β_m (λ) is given by the following relations:

$$\beta_m(\lambda) = A\lambda^{-4}$$

with

$$A = 1.09 * 10^{-3} \frac{P}{P_0} \frac{T_0}{T} \ (\text{km}^{-1} \mu\text{m}^4)$$

where

– P(mbar) is the atmospheric pressure and $P_0 = 1,013$ mbar.

– $T(K)$ is the atmospheric temperature and $T_0 = 273.15$ K.

Equation 5.3. *Molecular scattering coefficient*

As a result, scattering is negligible at infrared wavelengths. Rayleigh scattering primarily affects ultraviolet wavelengths up to visible wavelength. The blue color of the clear-sky background is due to this type of scattering.

5.3.3. *Aerosol absorption*

It results from the interaction between the radiated power and the aerosols, tiny particles suspended in the atmosphere (ice, dust, smoke, fog, and mist).

The absorption coefficient a_n is given by the following equation:

$$\alpha_n (\lambda) = 10^5 \int_0^{\infty} Q_a \left(\frac{2\pi r}{\lambda}, n'' \right) \pi r^2 \frac{dN(r)}{dr} dr$$

where

– $\alpha_n(\lambda)$ is the absorption coefficient due to aerosols (km^{-1}).

– λ is the wavelength (μm).

– $dN(r)/dr$ is the particle size distribution per unit of volume (cm^{-4}).

– n'' is the imaginary part of the refractive index n of the considered aerosol.

– r is the radius of the particles (cm).

– $Q_a(2\pi r/\lambda, n'')$ is the absorption cross section for a given type of aerosol.

Equation 5.4. *Molecular absorption coefficient*

Mie theory [MIE 08] predicts the electromagnetic field diffracted by homogeneous spherical particles. It assesses both physical quantities, such as the absorption normalized cross section Q_a and the scattering normalized cross section Q_d. They depend on the particle size, refractive index, and incident wavelength. They represent the section of an incident wave normalized by the geometric section of the particle (πr^2), such as the power absorbed (scattered) is equal to the power crossing this section.

The refractive index of aerosols depends on their chemical composition. It is a complex number and depends on the wavelength. It is noted $n = n' + n''$, where n' is a function of the scattering capacity of the particle and n'', is a function of the absorption of the same molecule.

Note that in the visible and near-infrared spectral region, the imaginary part of the refractive index is extremely low and can be neglected in the calculation of the global attenuation (extinction). In the far-infrared case, the imaginary part of the refractive index must be taken into account.

5.3.4. Aerosol scattering

It results from the interaction of light with particles (aerosols, hydrometeors) whose size is of the same order of magnitude as the wavelength.

The scattering coefficient β_n is given by the following equation:

$$\beta_n(\lambda) = 10^5 \int_0^\infty Q_d\left(\frac{2\pi r}{\lambda}, n'\right) \pi r^2 \frac{dN(r)}{dr} dr$$

where

– $\beta_n(\lambda)$ is the scattering coefficient due to aerosols (km^{-1}).

– λ is the wavelength (μm).

– $dN(r)/dr$ is the particle size distribution per unit of volume (cm^{-4}).

– n' is the real part of the refractive index n of the considered aerosol.

– r is the radius of the particles (cm).

– Q_d ($2\pi r/k)\lambda$, n') is the scattering cross section for a given type of aerosol.

Equation 5.5. *Aerosolar scattering coefficient*

The distribution of particle size is generally represented by an analytic function, such as the log-normal distribution for aerosols and the modified Gamma distribution for fog [BAT 92, KIM 01, KRU 62, NAB 04]. The latter is largely used to model different types of fog and clouds. It is given by the following equation [DEI 69, SHE 79]:

$$N(r) = ar^{\alpha} \exp(-br)$$

where

– $N(r)$ is the number of particles per unit volume whose radius ranges from r to $r + dr$,

– α, a, and b are the parameters that characterize the distribution of particle sizes.

Equation 5.6. *Distribution of particle size*

Software of atmospheric transmission calculations such as FASCOD, LOWTRAN, and MODTRAN considers two particular types of fog: the dense advection and the convection or moderate radiation which are modeled by the modified Gamma sizes distribution. Typical parameters are given in Table 5.1 [CLA 81, SHE 89].

Mie theory predicts the diffusion coefficient Q_d due to aerosols. It is calculated assuming that the particles are spherical and sufficiently distant from each other so that the scattered field by a particle and arriving on another particle can be calculated assuming far-field regime.

	α	a	b	N	W	r_m	V
Dense advection fog	3	0.027	0.3	20	0.37	10	130
Moderate radiation fog	6	607.5	3	200	0.02	2	450

Table 5.1. *Parameters characterizing the particles size distribution*

Notes:

– N is the total number of water particles per unit volume (nb/cm^3).

– r_m is the modal radius (μm) for which the distribution presents a maximum.

– W is the liquid water content (g/m³).

– V is the visibility associated to the fog type (m).

The scattering cross section Q_d is a function that strongly depends on the size of the aerosol compared to the wavelength. It reaches its maximum value (3.8) when the radius of the particle is equal to the wavelength: the scattering is then maximal. As the particle size increases, it stabilizes around a value equal to 2. We must therefore expect a very selective function by the particles whose radius is less than or equal to the wavelength. Clearly, scattering depends strongly on the wavelength.

Aerosol concentration, composition, and size distribution vary temporally and spatially; it is difficult to predict attenuation by these aerosols. Although their concentration is closely related to optical visibility, there is no unique particle size distribution for a given visibility. Visibility characterizes the transparency of the atmosphere as estimated by a human observer. It is measured by the RVR. The scattering coefficient is the most detrimental factor in terms of FSO wave propagation.

5.4. Models

Different models of attenuation exist in the literature: Kruse and Kim, Bataille, and Al Naboulsi models; rain and snow attenuations; and scintillations.

5.4.1. *Kruse and Kim models*

The attenuation coefficient for visible and near-infrared waves up to 2.4 µm is approximated by the following equation:

$$\gamma(\lambda) \cong \beta_n(\lambda) = \frac{3.912}{V} \left(\frac{\lambda_{nm}}{550} \right)^{-q}$$

where

– V is the visibility (km).

– λ_{nm} is the wavelength (nm).

The q coefficient characterizes the size particles. It is given by the following relation [KRU 62]:

$$q = \begin{cases} 1.6 & \text{if } V > 50 \text{ km} \\ 1.3 & \text{if } 6 \text{ km} > V > 50 \text{ km} \\ 0.528V^{1/3} & \text{if } V > 6 \text{ km} \end{cases}$$

As a result, attenuation is a decreasing function of the wavelength.

Equation 5.7. *Kruse and Kim model*

Recent studies have led to define the parameter q as follows [KIM 01]:

$$q = \begin{cases} 1.6 & \text{if } V > 50 \text{ km} \\ 1.3 & \text{if } 6 \text{ km} > V > 50 \text{ km} \\ 0.16V + 0.34 & \text{if } 1 \text{ km} > V > 6 \text{ km} \\ V - 0.5 & \text{if } 0.5 \text{ km} > V > 1 \text{ km} \\ 0 & \text{if } V < 0.5 \text{ km} \end{cases}$$

where V is the visibility (km).

As a result, attenuation is a decreasing function of the wavelength when visibility exceeds 500 m. For lower visibilities, the atmospheric attenuation is independent of the wavelength.

5.4.2. *Bataille's model*

Bataille's model [BAT 92] calculates the molecular and aerosol extinction for six laser lines (0.83, 1.06, 133, 1.54, 3.82, and 10.591 μm) by a polynomial approach on terrestrial links close to the ground. They are examined below.

5.4.2.1. *Molecular extinction*

The specific extinction coefficient σ_m is obtained by a 10-term expression:

$$\sigma_m = -\ln \begin{pmatrix} B_1 + B_2 T' + B_3 H + B_4 T' H \\ +B_5 T'^2 + B_6 H^2 + B_7 T' H^2 \\ +B_8 T'^2 H + B_9 H^3 + B_{10} T'^3 \end{pmatrix}$$

where

– $T' = T(K)/273.15$ is the reduced air temperature.

– H is the absolute humidity (g/m^3).

The coefficients B_i ($i = 1.10$) for different studied wavelengths are given in the literature [BAT 92] and [GEB 04].

5.4.2.2. Aerosol extinction

The specific extinction coefficient σ_n is obtained by a 10-term expression:

$$\sigma_n = -\ln \left(\begin{array}{l} A_1 + A_2 H + A_3 H^2 + A_4 H^x \\ + A_5 V^{-1/2} + A_6 V^{-y} + A_7 H V^{-1/2} \\ + A_8 (H/V)^y + A_9 H^z / V + A_{10} H V^{-1} \end{array} \right)$$

where

– V is the visibility (km).

– H is the absolute humidity (g/m^3).

– x, y, and z are real numbers used to optimize the polynomial for each studied wavelength; their values are adjusted when the maximum relative error between FASCOD and the polynomial is lower than 5%. The coefficients A_i ($i = 1.10$) for the different studied wavelengths are given in the literature [BAT 92, COJ 99] for both types of aerosol: rural and maritime.

5.4.3. Al Naboulsi's model

Al Naboulsi *et al.* [NAB 04] have developed simple relations from FASCOD enabling the attenuation in the range of wavelengths from 690 nm to 1,550 nm and visibility ranging from 50 m to 1,000 m for two types of fog: advection and convection.

Advection fog occurs when warm moist air moves over a cold floor. The air in contact with the ground cools and reaches its dew point, water vapor condenses. This type of fog occurs more particularly in spring when warm moist air moves from the south over cold regions. Attenuation by an advection fog is expressed by the following equation:

$$\sigma_{advection} = \frac{0.11478\lambda + 3.8367}{V}$$

where

– λ is the wavelength (μm).

– V is the visibility (km).

Equation 5.8. *Attenuation by an advection fog*

The radiation or convection fog is generated by the radiative cooling of an air mass due to the nocturnal radiation from the ground when meteorological conditions (very weak winds, high humidity, clear air, etc.) are favorable. Soil loses its heat accumulated during the day. It becomes cold. On contact, the air cools, reaches its dew point, and the humidity it contains condenses. We have the formation of a cloud close to the ground. This type of fog occurs most commonly in valleys. Attenuation by a convection fog is expressed by the following equation:

$$\sigma_{convection} = \frac{0.18126\lambda^2 + 0.13709\lambda + 3.7502}{V}$$

where

– λ is the wavelength (μm).

– V is the visibility (km).

Equation 5.9. *Attenuation by a convection fog*

5.4.4. Rain attenuation

Rain is formed from water vapor contained in the atmosphere. It consists of water droplets whose form and number vary in time and spatially. Their form depends on their size: they may be regarded as spheres until a radius of 1 mm and beyond as oblate spheroid: ellipsoid generated by the revolution of an ellipse around its minor axis. The equivalent radius is generally introduced, the radius of the sphere that has the same volume.

The rain attenuation (dB/km) is generally given by the Carbonneau's relation [CAR 98]:

$$Att_{rain} = 1.076 * R^{0.67}$$

Equation 5.10. *Rain attenuation*

Figure 5.2 shows the variations of the specific attenuation (dB/m) due to rainfall in the optical and infrared spectrum.

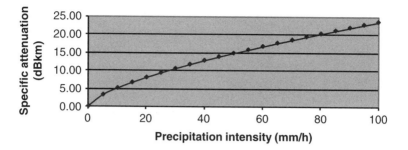

Figure 5.2. *Specific attenuation (dB/km) due to the rain*

Recommendation ITU-R P.837 gives the rain intensity R_p, exceeded for a given percentage of the average year p, and for any location [ITU 04].

5.4.5. *Snow attenuation*

Specific attenuation due to snow as a function of snowfall rate is given by the following equation:

$$Att_{snow}[dB/km] = aS^b$$

where

– Att_{snow} is the specific attenuation due to snow (dB/km).

– S is the snowfall rate (mm/h).

– a and b are functions of the wavelength (nm). They are given in Table 5.2.

Equation 5.11. *Attenuation due to snow*

	a	b
Wet snow	$0.0001023\lambda_{nm} + 3.7855466$	0.72
Dry snow	$0.0000542\lambda_{nm} + 5.4958776$	1.38

Table 5.2. *Parameters "a" and "b" for wet and dry snow*

The wet and dry snow attenuations as a function of snowfall rate for λ = 1,550 nm are given in Figures 5.3 and 5.4.

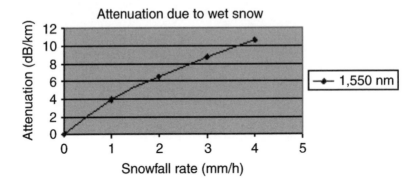

Figure 5.3. *Wet snow: attenuation in function of snowfall*

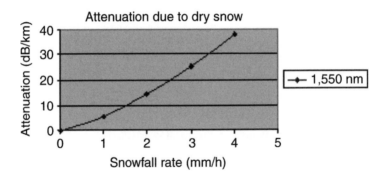

Figure 5.4. *Dry snow: attenuation in function of snowfall*

5.4.6. *Scintillation*

Randomly distributed cells of various sizes (10 cm to 1 km) and different temperature can be formed within the propagation medium under the influence of thermal turbulence. These different cells have different refractive indexes causing scattering, multiple paths, and variation in the angles of arrival: the received signal fluctuates rapidly at frequencies between 0.01 Hz and 200 Hz. The wave front varies similarly causing focusing and defocusing of the beam. Such fluctuations of the signal are called scintillation.

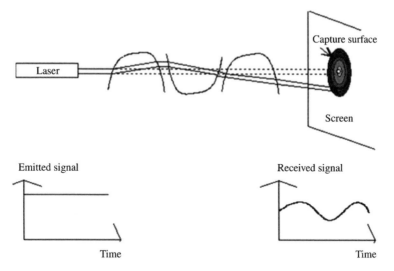

Figure 5.5. *Influence of a large turbulent cell (deviation)*

Figures 5.5–5.7 show schematically this effect as well as the variations (amplitude, frequency, etc.) of the received signal. The beam deviates when the heterogeneities are large compared to the beam cross section (Figure 5.5) and the beam is widened when heterogeneities are small (Figure 5.6). A mixture of heterogeneities results in scintillation (Figure 5.7) [WEI 08].

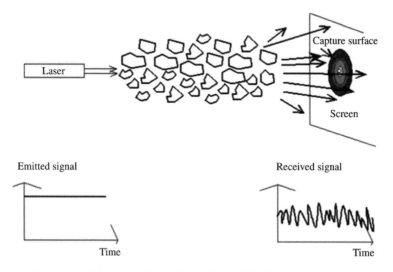

Figure 5.6. *Influence of a small turbulent cell (widening of the beam)*

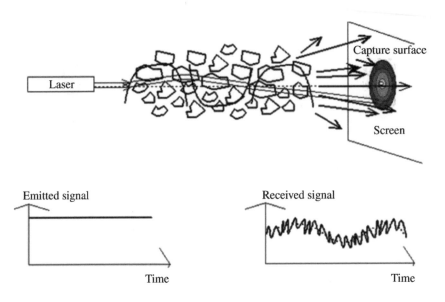

Figure 5.7. *Heterogeneities (scintillations)*

Tropospheric scintillation effects are usually studied from the logarithm of the amplitude χ [dB] of the observed signal (log-amplitude), defined as the ratio in decibels between the instantaneous amplitude and to its average value. The intensity and the fluctuations rate (scintillations frequency) increases with the frequency of the wave. For a plane wave, a weak turbulence, and a punctual receiver, the scintillation variance of "log-amplitude" σ_χ^2 [dB^2] can be expressed by the following relation:

$$\sigma_\chi^2 = 23,17 * k^{7/6} * C_n^2 * L^{11/6}$$

where

– k ($2\pi/\lambda$) is the wave number (m^{-1}).

– L is the length of the link (m).

– C_n^2 is the refractive index structure parameter, the turbulence intensity (m$^{-2/3}$).

Equation 5.12. *Scintillation variance*

The peak-to-peak amplitude scintillation is equal to $4\sigma_\chi$ and the attenuation due to the scintillation is equal to $2\sigma_\chi$. For strong turbulence, there is a saturation of the variance given by the above relation [BAT 92]. Note that the parameter C_n^2 has a different value at optical wave than at millimeter wave [VAS 97]. Millimeter waves are particularly sensitive to humidity fluctuation while at optical wavelength, the refractive index is essentially a function of temperature (the contribution of water vapor is negligible). At millimeter wavelength, C_n^2 value is approximately equal to 10^{-13} m$^{-2/3}$ which is a mean turbulent (usually in millimeter wavelength, we have $10^{-14} < C_n^2 < 10^{-12}$), and in optical wavelength, C_n^2 value is approximately equal to 2×10^{-15} m$^{-2/3}$ which is a weak turbulence (usually in optical wavelength, we have $10^{-16} < C_n^2 < 10^{-13}$) [BAT 92].

Figure 5.8 illustrates the variation of the attenuation of 1,550 nm wavelength optical beams for different types of turbulence over distances up to 2,000 m.

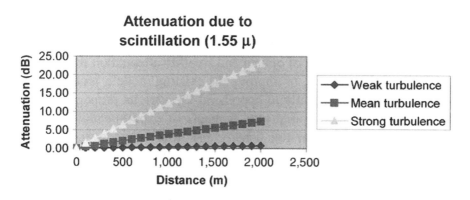

Figure 5.8. *Variation of the attenuation due to scintillation*

The international visibility code (Table 5.3) shows the attenuation (dB/km) for various climatic conditions [KIM 01]:

– weather conditions (from very clear periods to dense fog);

– precipitation (mm/h): drizzle, rain, storm;

– visibility from 50 km to 50 m.

International visibility code				
Weather conditions	**Precipitation**	**mm/h**	**Visibility (m)**	**Attenuation (dB/km)**
Dense fog			0	
			50	315
Thick fog			200	75
Moderate fog			500	28.9
Light fog	Storm	100	770	18.3
Very light fog			1,000	13.8
	Strong rain	25	1,900	6.9
	Snow		2,000	6.6
Light mist	Average rain	12.5	2,800	4.6
			4,000	3.1
Very light mist	Light rain	2.5	5,900	2
			10,000	1.1
Clear air	Drizzle	0.25	18,100	0.6
			20,000	0.54
Very clear air			23,000	0.47
			50,000	0.19

Table 5.3. *International visibility code*

5.5. Experimental set-up

We describe below the experimental set-up carried out on the site of La Turbie (France) to characterize the optical propagation channel in the presence of different meteorological conditions (rain, hail, snow, fog, mist, etc.). The aim is multiple [NAB 04]:

– to study the influence of the atmosphere on the propagation of a laser beam (FSO);

– to compare atmospheric attenuation models against carried measurements;

– to determine the most reliable and most realistic attenuation model due to fog;

– to define the most robust wavelength.

It consists of:

– an FSO link operating at 650, 850, and 1,550 nm;

– a meteorological station;

– a transmissometer.

Figure 5.9 represents a synoptic view of the experimental set-up.

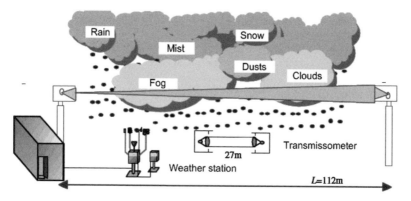

Figure 5.9. *Synoptic view of the experimental set-up*

Figure 5.10 shows a picture of the transmissometer (narrow vertical beam) and of the 650 nm laser (divergent horizontal beam) light beam taken at night in the presence of dense fog.

Figure 5.10. *Light beam of the transmissometer and laser*

The implemented weather station is composed of the following sensors:

– a thermometer (outside temperature);

– a barometer (atmospheric pressure);

– a hygrometer (atmospheric humidity);

– an anemometer (instantaneous and average wind speed);

– a weather vane (wind direction);

– a rain gauge (rainfall);

– a pyranometer (solar radiation).

5.6. Experimental results

We present some experimental results (Figures 5.11 and 5.12) deduced from the measurements of attenuation in function of the visibility carried out under the project COST 270 in cooperation with the University of Graz (Austria) [GEB 04].

5.6.1. *Comparaison with Kruse and Kim model (850 nm)*

Figures 5.11 and 5.12 show the evolution of the measured attenuation at the site of La Turbie of the specific attenuation (dB/km) of the light beam at 850 nm in function of the visibility in the presence of fog.

The results are compared with Kruse and Kim models.

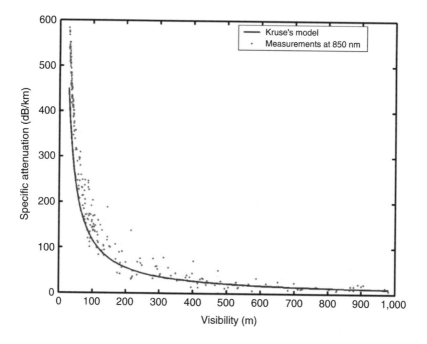

Figure 5.11. *Measurements and comparison with Kruse's model*

5.6.2. *Comparaison with Al Naboulsi's model*

Figures 5.13 and 5.14 show the evolution of the measured attenuation at the site of La Turbie of specific attenuation (dB/km) of the light beam at 850 nm in function of the visibility in the presence of fog.

The results are compared with Al Naboulsi's models (advection and convection).

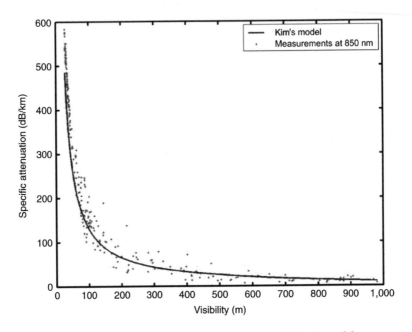

Figure 5.12. *Measurements and comparison with Kim's model*

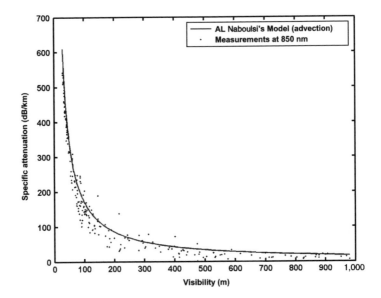

Figure 5.13. *Measurements and comparison against Al Naboulsi's model (advection)*

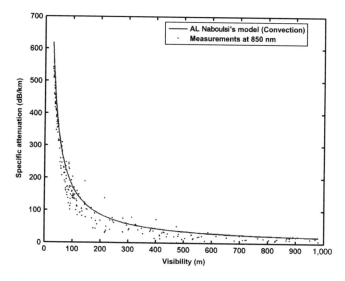

Figure 5.14. *Measurements and comparison against Al Naboulsi's model (convection)*

Measurement comparisons to existing models in the literature shows a good agreement between measurements and models available.

From analysis of the previous curves, it appears that the Al Naboulsi model, developed from FASCOD, is in excellent agreement with experimental measurements for low visibility while Kruse and Kim models deviate significantly from the measurements.

5.7. Fog, haze and mist

Fog, haze, and mist are composed of fine water droplets (<100 µm) suspended in the air. In some very cold countries, they can be particles of ice. It is formed by a process of condensation near the ground resulting from a radiative cooling of the earth (especially at night) or the passage of air on a cold ground (advection fog or advection mist).

To distinguish fog from haze or mist, it is generally agreed that the visibility is greater than 1,000 m in the presence of haze while it is less than 1,000 m in the presence of fog. Their compositions and their size distributions vary widely. In general, fogs have a water content lower than clouds (<0.1 gm^{-3}) and a smaller concentration of droplets (<100 cm^{-3}). It is characterized by visibility, determined by the maximum distance beyond which a prominent object cannot be seen by a human observer. The occurrence of fog is very variable depending on location, proximity to the sea or a large lake, mountainous zones, etc.

The liquid water concentration in the fog is typically equal to about 0.05 g/m^3 for a moderate fog (visibility of about 300 m) and 0.5 g/m^3 for dense fog (visibility of about 50 m) [ITU 05].

NOTE:– Meteorological measurements carried out in Belgium on 16 stations in 20 years provide information on its maximum (worst case) and median (not exceeded in 50% of cases) annual frequency. Table 5.4 gives the annual frequency of these fog events [BOD 77].

	Median frequency (%)	Maximal frequency (%)
Light, moderate, or thick fog ($V < 1,000$ m)	0.055	0.14
Moderate or thick fog ($V < 500$ m)	0.035	0.11
Thick fog ($V < 200$ m)	0.020	0.08

Table 5.4. *Annual frequency of fog events in Belgium*

Figure 5.15 shows the distribution in France of the number of days per year with the presence of fog (days during which there is, even temporarily, a reduction in visibility to less than 1 km).

5.8. The runway visual range (RVR)

5.8.1. *The visibility*

Originally defined for the purposes of meteorology, the RVR (or visibility) is the distance for which a parallel luminous ray beam, generated by an incandescent lamp, at a color temperature of 2,700 K, must cover so that the luminous flux intensity is reduced to 0.05 of its initial value. It characterizes the transparency of the atmosphere.

Figure 5.16 provides an example of changes in the RVR observed at the site of La Turbie (06) 28 June 2004 during a day of low visibility (<10,000 m) and in the presence of fog (<1,000 m).

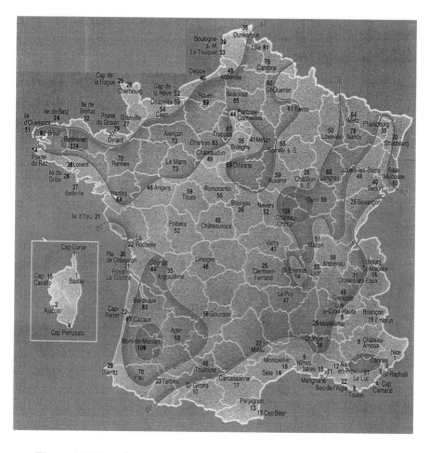

Figure 5.15. *Distribution in France of the number of days with fog per year*

It is measured using a transmissometer or a scatterometer. The transmissometer is an instrument based on the loss of light intensity of a beam of light rays in the atmosphere, which depends on both absorption and diffusion.

The scatterometer gives an indication of the visibility in the atmosphere by measuring the diffusion of a beam light from a given volume.

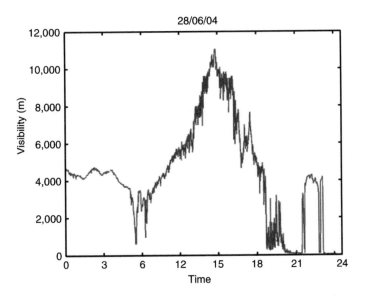

Figure 5.16. *RVR variations observed on the site of La Turbie*

5.8.2. Measuring instruments

5.8.2.1. *The transmissometer*

The transmissometric method is the most commonly used to measure the average extinction coefficient in an horizontal air cylinder placed between a transmitter with a modulated light source and a receiver with a photodetector (usually a photodiode located at the parabolic mirror or a lens focal point).

The halogen lamp or the light discharge tube in the xenon is the most commonly used light source. The modulation of the light source avoids the influence of the solar flare.

The current from the photodetector determines the transmission factor which is used to calculate the extinction coefficient and the RVR.

There are two types of transmissometer [COJ 99]:

– Those where the transmitter and the receiver are placed in different cases and at a known distance from each other (Figure 5.17).

– Those where the transmitter and the receiver are placed in the same case, the emitted light is reflected by a remotely placed mirror or retro reflector (Figure 5.18).

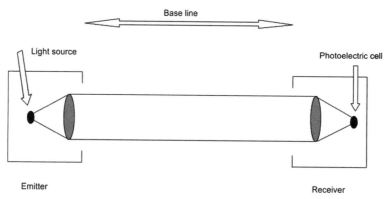

Figure 5.17. *Direct beam transmissometer*

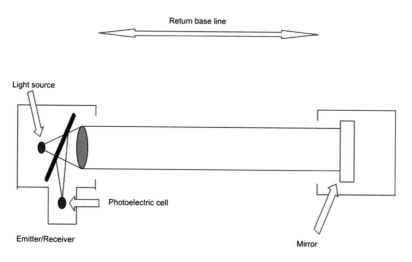

Figure 5.18. *Reflected beam transmissometer*

The distance covered by the light between the transmitter and the receiver is commonly called "the transmissometer base" and can vary from a few meters up to 300 m. Figure 5.19 shows the emission part of a transmissometer installed on the site of La Turbie during the experiment carried out to study the effects of fog on the visible and infrared propagation of FSO in the Earth's atmosphere.

Figure 5.19. *Emission part of a transmissometer installed on the site of La Turbie*

5.8.2.2. *The scatterometer*

The most convenient method to perform this measurement is to focus a light beam on a small volume of air and determining, by photometric means, the proportion of light scattered in a large enough solid angle and in not preferred directions.

Two types of measurement are used in these instruments: backscatter and forward scatter [COJ 99].

Backscatter (Figure 5.20): The light beam is concentrated in a small volume of air; it is backscattered and collected by the photoelectric cell.

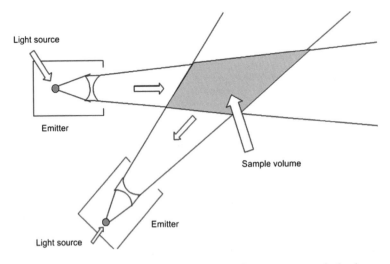
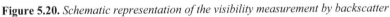

Figure 5.20. *Schematic representation of the visibility measurement by backscatter*

Forward scatter: The instruments consist of a transmitter and a receiver whose emission and reception beams form between them an angle between 20° and 50° (Figure 5.21). Other devices place a diaphragm halfway between the transmitter and the receiver, or two diaphragms placed close to the transmitter and the receiver.

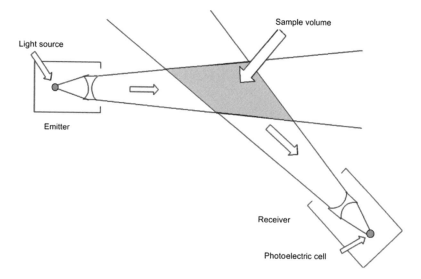

Figure 5.21. *Schematic representation of the measurement of forward scatter visibility*

Figure 5.22. *Example of a scatterometer implemented on a motorway area*

Figure 5.22 shows a photo of a scatterometer implemented on a motorway area to measure the fog intensity and announce to motorists the atmospheric visibility.

The knowledge of the visibility finds numerous applications in meteorology (identification of an air mass for synoptic meteorology and climatology), in aeronautics (determination of RVR), in telecommunications (evaluation of the effects of atmospheric particles (fog and aerosols) on optical, visible, and near-infrared transmission until 2.5 μm, determination of FSO links in presence of fog more particularly), and in terrestrial or maritime traffic security (measurement of the visibility in the fog) [DEF 07, DEF 10, HAU 05, HAU 06].

5.9. Calculating process of an FSO link availability

Software simulating FSO link in terms of probability of availability or interruption (QoS) has been developed at Orange Labs [ORA 11, CHA 05]. It is a decision support tool for the development of point-to-point high-rate FSO links over short distances. It includes data on the equipment (power, wavelength, sensitivity, etc.), the location of a site (coordinates, altitude, height/ground, etc.), and climatic and atmospheric parameters (relative humidity, soil roughness, albedo, solar radiation, etc.). It relies on the available technical knowledge relative to the effects of aerosols, scintillation, ambient light, rain, etc., the proprietary algorithms to model impairments due to snow or visibilities below 1 km, and the important phenomena such as advection and convection fog, weather statistics, and experiments analysis results. Applied to an experimental link, the results derived from the software showed excellent agreement between the predicted and the effective availability [NAB 05].

Three screens characterize the software: input data acquisition, results, and link profile presentation.

– The input data acquisition screen allows us to enter the following information (Figure 5.23):

– the data defining the link physical location (latitude, longitude, altitude, height above ground, orientation, etc.);

– the data characterizing the used equipment (number of transmitters, power, diameter, divergence, etc.);

– the data providing environmental information of the site (relative humidity, albedo (percentage of reflected solar energy), roughness, solar radiation, distance, etc.);

– the equipment data (wavelength, rate, system loss, etc.);

– the environment (urban, rural, marine).

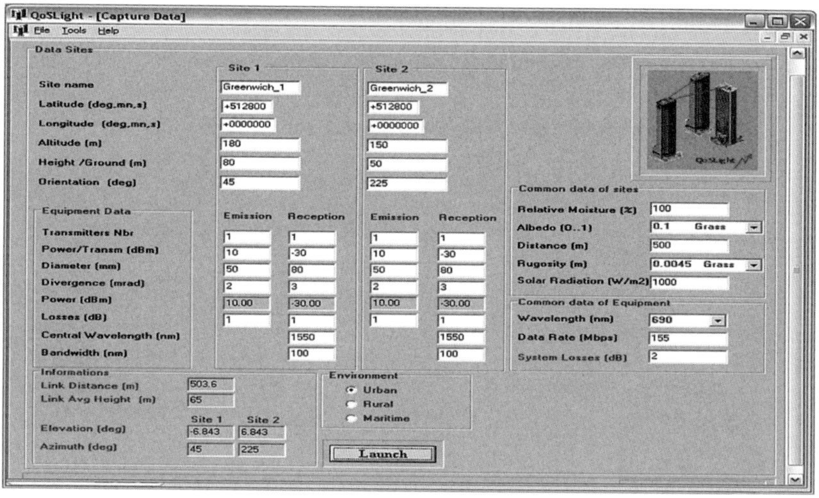

Figure 5.23. *Input data acquisition screen*

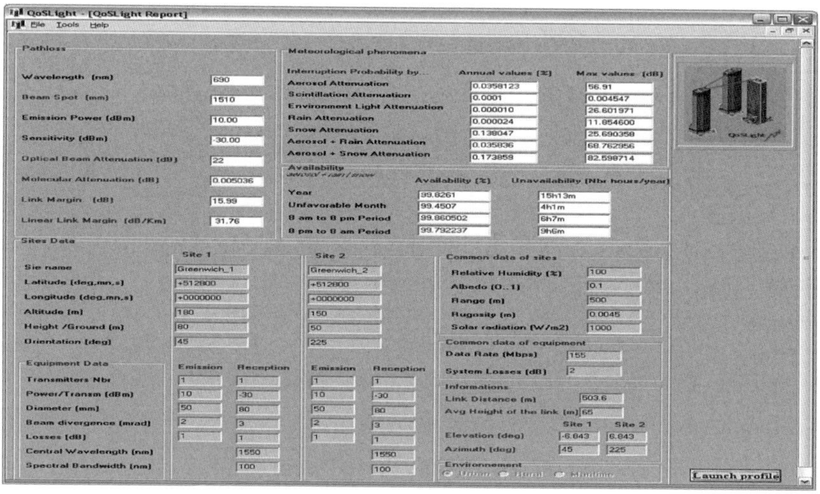

Figure 5.24. *Results presentation screen*

The results presentation screen displays the following results (Figure 5.24):

– the link information (length, medium height, elevation, azimuth, etc.);

– weather information;

– the data sites (equipment, environment, etc.);

– the link availability and the interruption probability.

The profile screen shows the wave front propagation (Figure 5.25).

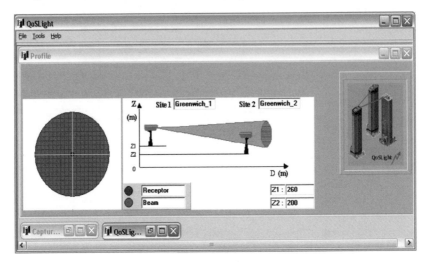

Figure 5.25. *Screen showing the wave front propagation*

5.10. Conclusion

The various aspects of the photon propagation in the atmosphere have been presented (molecular and aerosol absorption, molecular and aerosol diffusion (hail, smoke, haze, fog), rain and snow attenuation, and scintillation effects). They are the key to any good understanding of future communication systems using wireless optical links. Fog seems to be the most detrimental element to the operation of FSO links.

The comparison of experimental data allowed us to validate propagation models proposed in the literature. These are used to control transmission power levels of future FSO links by ensuring a sufficient momentum taking into account the variability of optical propagation conditions.

Experimental links showed that the FSO is a reliable broadband alternative to laying optic fiber and leads to greater acceptance of this technology in the industry of high-rate telecommunications networks.

To better understand the availability of an FSO link, the reader is referred to quality of service simulation tools. These tools allow us, for a given geographic site, to determine the availability and reliability of a link based on system parameters (power, wavelength, and equipment characteristics) and climatic and atmospheric parameters. They integrate the various physical phenomena responsible for the link blackout, such as attenuation due to ambient light, scintillation, rain, snow, and fog. Practical examples are provided in Chapter 11.

These various elements described in this chapter have contributed to the development of new recommendations at International Telecommunication Union (ITU), dedicated to propagation data and prediction methods required for the design of terrestrial FSO links [ITU 03, ITU 07a, ITU 07b].

The laser beam propagation in the atmosphere has been studied extensively in recent years. Many items are available in the literature. The reader can refer to them in [LEI 03, WAI 05, GRA 07, POP 09, WAL 09, STO 09, HEN 10].

Chapter 6

Indoor Optic Link Budget

"Light is an old friend that has fascinated me
ever since I can remember"
Jeff Hecht,
Understanding Fiber Optics,
Prentice Hall, Upper Saddle River, 2002

6.1. Emission and reception parameters

For indoor (propagation limited to contours of a room) or outdoor communication, a detailed analysis of the assessment of connection (link budget) is a very significant step for any system. For example, for the systems with optical fibers, the engineer examines the power which is injected into fiber by the transmitter and then determines all the potential losses and gains until the signal arrives on the reception device. The receiver typically has a specific sensitivity (S_e) to a given throughput. This sensitivity represents the lower limit of the optical level of power rather than the reception for a quality of preset transmission. To guarantee the required quality of service (QoS), it is necessary to make sure that after having withdrawn all the losses, the received power remains above the sensitivity of the receiver. To make the connection more resistant, it is judicious to work on a level of power in reception slightly higher than the necessary minimum capacity: the variation is called margin and constitutes a reserve of power used in the event of degradation of the channel. This step represents the compromise between the distance from transmission and the reliability of the communication.

Like the devices in the field of radio or fiber optics, this process is similar for optical wireless systems.

In general, the principal parameters to be taken into account are:

– optical power transmitted by source (P_t), whose values range from 0 (1 mW) to 30 (1 W) dBm according to the wavelength and cost of the emission module;

– the divergence of the emission beam (DIV), whose total value should be of approximately 90° or ±45°;

– the sensitivity of the receiver (S_e), whose values range from –90 (1 nW) to –20 (10 μw) dBm according to the throughput and the detection surface;

– the effective surface of reception (A_{eff}), which generally includes the active surface of the receptor and an optical antenna of concentration;

– the aperture of receptor (field of view – FOV), whose total value should also be of approximately 90° or ±45°;

– the positions and the distance from the couple transmitter/receiver.

NOTE.– The reader will find further information relating to the elements of geometrical optics, photometry, and energy, and to the logarithmic notation, respectively, in Appendices 1 and 2.

Before reaching the receptor, the optical beam is attenuated. To understand how part of the transmitted power arrives at the receptor, it is necessary to examine the various losses in the wireless optical channel. The losses to be considered are:

The optical losses: This is mainly due to the imperfect lenses and the losses due to interfaces. This attenuation is generally integrated in the value of transmitted power (P_t) or the sensitivity (S_e).

The loss of pointing: For indoor environments, these effects are supposed to be negligible.

The atmospheric loss: In the event of significant atmospheric scintillation, it is possible to have a degradation of the signal. But the distance considered in an interior environment makes this attenuation negligible.

The molecular loss: This represents the attenuation by the molecules of the atmosphere. This attenuation is a function of the used wavelength. But their linear values are sufficiently low (maximum 0.41 dB/km at 850 nm) to be also regarded as negligible for indoor environment.

The geometrical loss: This attenuation is due mainly to the divergence of the optical beam, the distance, and the effective reception surface. The geometrical attenuation is the most significant loss. For example, with identical parameters of P_t, S_e, and A_{eff}, therefore the same value of geometrical attenuation, then three values of divergence (DIV equivalent to ±1°, ±15°, and ±80°); Figure 6.1 shows the various values of the maximum distance on the normal axis of connection.

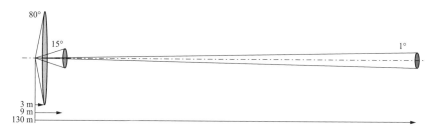

Figure 6.1. *Distance as a function of the divergence*

The maximum distance varies very quickly with the value of the divergence. For a given value of the divergence, when the distance is doubled, the geometrical attenuation is quadrupled. A low divergence is generally used in the point-to-point outside communications (free-space optics – FSO) whereas a strong divergence is used for the remote controls.

6.1.1. *Transmission device: parameters*

One of the significant points relates to the calculation of the emitted optical power, and the way in which this power is connected to the light intensity, quantity varying according to the direction. The emitted optical power is the integral of the light intensity on all the directions of emission. The intensity of its radiation $I(\varphi, \beta)$ is defined like the power per solid angle unit emitted by the source with the position (φ, β) compared to its normal.

This is generally considered the case for a source, as shown in Figure 6.2, such as the power, which is radiated according to a circular symmetry around the vertical.

This figure represents the elements of the calculation of the optical power emitted on a solid angle equal to 2π; the surface of the shaded zone is equal to $d\Omega$.

Owing to this symmetry of revolution around its optical axis, the intensity of the source depends only on one Φ angle. We can then model it according to a generalized Lambert model [GFE 79, BAR 93].

Indeed, to calculate the optical power emitted on a solid angle equal to 2π using such a source having a symmetry of revolution around its optical axis, we cut out the solid angle Ω in a succession of annular angles elementary $d\Omega$ ranging between the angles Φ and $\Phi + d\Phi$:

$$d\Omega = 2\pi \sin(\Phi) d\Phi$$

Equation 6.1. *Elementary annular angle*

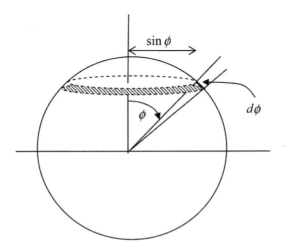

Figure 6.2. *Schematization of the elements of calculation of the optical power*

If the radiated power follows the generalized Lambert model, we will have the following equation:

$$I(\Phi) = I(0)(\cos(\Phi))^m$$

Equation 6.2. *Light intensity*

Therefore, radiated total power P_t will be:

$$P_t = \int_0^{\pi/2} I(\phi) 2\pi \sin(\phi) d\phi = \int_0^{\pi/2} I(0)(\cos \phi)^m 2\pi \sin(\phi) d\phi$$

$$P_t = 2\pi I(0) \int_0^{\pi/2} (\cos\phi)^m \sin(\phi)d\phi$$

$$P_t = 2\pi I(0) \int_0^1 u^m du = 2\pi I(0)\frac{1}{m+1}$$

Equation 6.3. *Radiated total power*

And the normal intensity radiated in the normal direction with the source of light is given by:

$$I(0) = \frac{m+1}{2\pi}P_t$$

Equation 6.4. *Intensity radiated in the normal direction*

The device of emission will thus be characterized by two significant elements:

– the transmitted average optical power P_t (in mW or dBm);

– the half-power (HP) angle (in degrees) (optical transmitted HP angle).

This angle HP, such as the light intensity, lowers half and it indicates that the source can light uniformly in all the directions (Figure 6.3):

$$I(HP) = I(0)\cos(HP)^m = \frac{I(0)}{2}$$

Equation 6.5. *Determination of the angle HP*

From this angle HP, it is possible to determine the parameter m:

$$\cos(HP)^m = \frac{1}{2} \Rightarrow m = \log(1/2)/\log(\cos(HP))$$

Equation 6.6. *Determination of the value m*

Table 6.1 presents some m values for various values of HP for a generalized Lambert model.

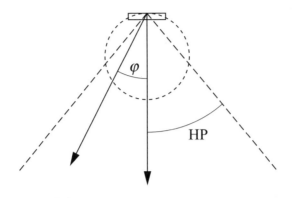

Figure 6.3. *Emission diagram and half-power angle*

HP(°)	10	15	20	30	40	45	50	60	70	80
m	45.28	20	11.14	4.82	2.6	2	1.57	1	0.646	0.396

Table 6.1. *Half-power angles and their respective m parameters*

In the case of a device with several systems of emission coupled, the ones with the others, in order to increase the divergence, we can consider, initially, an average value of total power and an HP angle including all the systems of emission.

From equations [6.2] and [6.4], it becomes possible to calculate the power radiated per solid unit of angle (φ):

$$P(\varphi) = \frac{m+1}{2\pi} P_t (\cos \varphi)^m \text{ with } \varphi \in [0, \pi/2]$$

Equation 6.7. *P(φ) power*

Figure 6.4 shows an example of emission diagram with HP equal to 30°, then a divergence (DIV) at ±30°.

With an aim of obtaining a significant HP angle, it is possible to use optical diffusers in front of the source of emission, but this increase also gives a significant geometrical attenuation. The use of several sources (multisector) in three dimensions makes it possible to improve the total HP angle and the emission coverage.

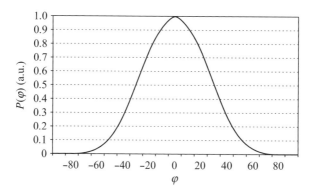

Figure 6.4. *Example of emission diagram*

6.1.2. *Reception device*

An optical reception device is made up mainly of an optical system to collect and concentrate the incident radiation, an optical filter making it possible to reject the optical disturbers, and a photodiode that converts the light intensity into electrical current. After these optic and electro-optic elements, there are additional amplification modules and data recovery in the electric field.

In complement of the sensitivity of the reception device expressed in mW or dBm, the effectiveness of the receptor is determined by two other significant elements: the effective surface of collection A_{eff} (mm^2) and the aperture or the FOV corresponding to the direction exposed to 50% of the maximum of received intensity.

Figure 6.5 shows an example of a receptor (R_x) placed at a distance d of the emission source (T_x).

The optical device of reception (R_x) detects an optical power that is directly proportional to its effective light-collection area. Increasing the photodiode area is expensive, and tends to decrease receiver bandwidth and increase receiver noise. Hence, it is desirable to employ an optical concentrator to increase the effective area. Concentrators may be imaging or of the non-imaging variety. The imaging concentrator reproduces the image at the image focal point. The telescopes used in long-range, free-space optical links represent examples of imaging concentrators. The non-imaging concentrators are used for most short-range links. The majority of the indoor wireless optical systems are carried out with a non-imaging concentrator, such as a hemispherical or aspheric lens. The aspheric lens is a lens close to a portion of sphere but not strictly spherical.

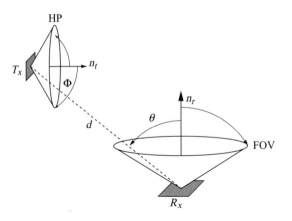

Figure 6.5. *Example of emission (T_x) and reception (R_x) devices*

For an incidence angle θ compared to the normal axis of receiver (R_x), an ideal non-imaging concentrator that has an index of refraction n has a gain of $g(\theta)$ [BAR 94]:

$$g(\theta) = \begin{cases} \dfrac{n^2}{\sin(FOV)^2}, & 0 \leq \theta \leq FOV \\ 0, & \theta > FOV \end{cases}$$

Equation 6.8. *Gain in the case of a non-imaging concentrator*

In general, an optical filter is associated in order to reject the undesirable wavelengths and attenuate the ambient light as much as possible [KAH 97]. Their transfer functions $T_f(\theta)$ depend on the angle of incidence. The value of the effective surface of collection of an optical receptor device (R_x), having a photodiode, an optical filter, and a concentrator, is given by the following equation:

$$A_{eff}(\theta) = \begin{cases} A_{ph}\cos(\theta)T_f(\theta)g(\theta), & 0 \leq \theta \leq FOV \\ 0, & \theta > FOV \end{cases}$$

Equation 6.9. *Effective surface (1st representation)*

The non-imaging concentrator represents the optimal solution between the effective surface and the aperture (FOV). By supposing an ideal optical filter with a transfer function corresponding to the unit, equation [6.9] can be rewritten as:

$$A_{eff}(\theta) = A_{ph}\cos(\theta)\frac{n^2}{\sin(FOV)^2}rect\left(\frac{\theta}{FOV}\right) = A_{eff}\cos(\theta)$$

Equation 6.10. *Effective surface (2nd representation)*

where

– A_{eff} is the maximum effective surface corresponding to a transmission with the normal (n_r) of the device of reception (R_x).

– rect (.) is a rectangular function defined by:

$$rect(x) = \begin{cases} 1 \text{ if } |x| \leq 1 \\ 0 \text{ if } |x| > 1 \end{cases}$$

Equation 6.11. *Rectangular function*

Beyond a certain angular value, the incident beam does not reach any more the active surface of the photodiode, as a result the communication is not carried out any more. This following relation is deduced from equation [6.10]:

$$A_{eff}(\theta)\sin^2(FOV) \leq n^2 A_{ph}$$

Equation 6.12. *Limit angle of reception*

This relation gives us the effective limit of collection surface available according to an FOV and a surface aperture of photodiode. An increase in the effective surface of collection induces an increase in the received power, but it reduces the FOV and thus the system coverage. A compromise is thus essential in the systems design.

The last significant parameter is the sensitivity of the optical receptor (S_e). It is the minimal value of received optical power allowing a correct detection of the transmitted signal. The sensitivity S_e corresponds to the detection threshold based on the dark current (I_d) of the photodiode. When the value of the optical power received on the active surface of the photodiode is lower than this sensitivity, the photodiode does not produce any answer. The sensitivity S_e is proportional to the responsivity

(R), which is expressed in Ampere/Watt, and the dark current (I_d) in Ampere, and is given by the following expression:

$$S_e = 10 \times \log\left(\frac{I_d/R}{0.001}\right)$$

Equation 6.13. *Theoretical sensitivity of the photodiode*

The sensitivity S_e is expressed in dBm or mW. It depends mainly on the bandwidth (which is related to the active surface) and the photodiode type. For example, an APD photodiode with a spectral response (R) of 0.43 A/W at 830 nm and a dark current (I_d) of 0.5 nA has a theoretical sensitivity of −59 dBm.

Actually, the sensitivity of the receptor must take into account the losses of internal optics of the optical reception device, such as the concentrator or the filter. Moreover, the sensitivity of the reception device must also take into account the perturbation due to the ambient light, the thermal noise, the electric noises, and the losses in the electric circuits, such as the amplifiers. Thus, the value of sensitivity of an optical reception device including an APD, as mentioned above, becomes −36 dBm.

6.2. Link budget for line of sight communication

6.2.1. *Geometrical attenuation*

We describe a generic model of calculation for a line of sight (LOS) and wide line of sight (WLOS) link budget, allowing us to consider the devices as black boxes. This approach makes it possible to integrate the devices characteristics of filtering, concentration, optical divergence, and the losses of the emission and reception devices. We indicated that the most significant attenuation is the geometrical attenuation; therefore, this value is related to the transmitted average power (P_t) and the power received (P_r) by the following relation:

$$P_{r(dB)} = P_{t(dB)} - Aff_{geo(dB)}$$

where $Aff_{geo} = 10\log H(0)$ and $H(0)$ is the channel gain at the null frequency or the attenuation of the useful average optical power.

Equation 6.14. *Relation between P_t and P_r*

For the LOS and the WLOS having only one path, it is possible to calculate only $H(0)$. Let us have a device of emission (Figure 6.5) whose emitted power is identical to generalized Lambert model. If we suppose that the value of the effective surface of the reception device (A_{eff}) is very low compared to the square of the distance from the couple transmitter/receptor ($A_{eff} << d^2$), then it is possible to write the received average optical power as the following relation (the Φ angle is the angle between the normal axis of the source (n_t) and the direct path):

$$P_r = I(\Phi) d\Omega$$

where

– $d\Omega = \cos(\theta) A_{eff} / d^2$ is the solid angle (Sr), when the receiver is seen since the source,

– θ is the angle between the normal axis (n_r) of the receptor and the direct path,

– $I(\Phi)$ is the radiant intensity (W/Sr).

Equation 6.15. *Relation between P_r and I*

From relations [6.7] and [6.15], we obtain:

$$P_r = \frac{m+1}{2\pi} P_t \cos(\Phi)^m \cos(\theta) \frac{A_{eff}}{d^2}$$

Equation 6.16. *Received power*

Consequently, for an LOS or a WLOS, the channel gain (direct current) or the linear geometrical loss is given by:

$$H(0) = \frac{m+1}{2\pi} \cos(\Phi)^m \cos(\theta) \frac{A_{eff}}{d^2}$$

Equation 6.17. *Linear geometrical loss in line of sight*

The corresponding impulse response can be comparable with a simple filter defined by [BAR 93]:

$$h(t) = \frac{m+1}{2\pi} \cos(\Phi)^m \cos(\theta) \frac{A_{eff}}{d^2} rect(\theta / FOV) \delta(t - d/c)$$

where c is the speed of light.

Equation 6.18. *Response of the impulse of the channel*

6.2.2. *Optical margin*

The optical margin of a connection M_l (dB) is the difference between the power available in reception and the receptor sensitivity:

$$M_l = P_r - S_e$$

Equation 6.19. *Margin of the system*

It is pointed out that a relatively high margin implies a high reliability of a wireless optical communication system, but that is with the detriment of the connection distance. Indeed, to assign most of the power of the signal to the margin of the system induces a reduction in power available in LOS with the distance. In fact, it is advisable to have a compromise between the distance and the reliability.

6.2.3. *Coverage*

Within the framework of a connection in LOS or WLOS, the coverage surface (A_c) can be defined for an angular value corresponding to HP (at −3dB or 50%). It can be expressed in the following way (in m^2):

$$A_c = \pi.(d.\tan HP)^2$$

where d is the transmitter–receiver distance (m).

Equation 6.20. *Coverage surface*

For example, the coverage surface of an emission device, with an HP angle of 30° and a distance d of 2 m, will be 4.2 m^2.

6.2.4. *Reciprocity and not reciprocity of the channel*

An indoor wireless optical communication is symmetrical when the emission and reception apparatuses are identical and have the same optical characteristics, and the characteristics of the uplink and the downlink are similar. However, unlike the radio, it is generally supposed that a base station, in a part, communicating with one or more modules will not have the same parameters with respect to the modules. The parameters of the uplinks and downlinks are not equal any more. To illustrate this, we suppose a base station (*BS*) and a module (*Mo*) in a topology in LOS or WLOS.

The emission–reception devices of the base station (*BS*) are characterized by:

– an emission power P_t^{BS}, for the downlink;

– its radiant intensity $I^{BS}\left(\Phi^{BS}\right)$, where Φ^{BS} is the angle of emission compared to the axis normal of the *BS*;

– the effective surface of collection $A_{eff}^{BS}\left(\theta^{BS}\right) = A_{eff}^{BS}\cos\left(\theta^{BS}\right)$, where θ^{BS} is the angle of reception compared to the axis normal of the *BS* and A_{eff}^{BS} has the maximum value when the angle θ^{BS} is null;

– the FOV FOV^{BS} of the uplink.

The module (*Mo*) is also characterized by:

– an emission power P_t^{Mo} for the downlink;

– its radiant intensity $I^{Mo}\left(\Phi^{Mo}\right)$, where Φ^{Mo} is the angle of emission compared to the axis normal of the module;

– the effective surface of collection $A_{eff}^{Mo}\left(\theta^{Mo}\right) = A_{eff}^{Mo}\cos\left(\theta^{Mo}\right)$, where θ^{Mo} is the angle of reception compared to the axis normal of the module and A_{eff}^{Mo} has the maximum value when the angle θ^{Mo} is null;

– the FOV FOV^{Mo} of the downlink.

When dimensions of the base station and the module are supposed as small as possible, the distance *d* is relatively identical for the downlink and the uplink in a given configuration. We obtain:

– for the downlink:

$$\mathrm{BS}\left\{ P_t^{BS}, I^{BS}\left(\Phi^{BS}\right) \right\} \mapsto \mathrm{Mo}\left\{ FOV^{Tm}, A_{eff}^{Tm}, \theta^{Tm} \right\}$$

– for the uplink:

$$\mathrm{Mo}\left\{ P_t^{Tm}, I^{Tm}\left(\Phi^{Tm}\right) \right\} \mapsto \mathrm{BS}\left\{ FOV^{BS}, A_{eff}^{BS}, \theta^{BS} \right\}$$

Based on equation [6.18], the gain will be thus (equivalent to the loss of propagation between transmitter and receiver):

$$H(0)^D = \frac{I^{BS}\left(\Phi^{BS}\right)}{P_t^{BS}} \frac{A_{eff}^{Mo}}{d^2} \cos\left(\theta^{Tm}\right) = \frac{m^{BS}+1}{2\pi} \cos\left(\Phi^{BS}\right)^m \frac{A_{eff}^{Mo}}{d^2} \cos\left(\theta^{Mo}\right)$$

$$H(0)^M = \frac{I^{Mo}\left(\Phi^{Mo}\right)}{P_t^{Mo}} \frac{A_{eff}^{BS}}{d^2} \cos\left(\theta^{BS}\right) = \frac{m^{Mo}+1}{2\pi} \cos\left(\Phi^{Mo}\right)^m \frac{A_{eff}^{BS}}{d^2} \cos\left(\theta^{BS}\right)$$

Equation 6.21. *Gain DC of the channel*

in which m^{BS} and m^{Mo} are, respectively, the numbers of mode from the base station and the module determined by their HP angle.

In diffusion mode (DIF) propagation, it is advisable to take into account the gains for each path. For indoor wireless optical communication, the channel gains of the downlink and of the uplink are generally different because of this asymmetry. The link budget evaluation must be calculated in two directions.

In the radio field, one antenna can be used for transmission and reception. In this configuration, the two responses are identical and the channel is reciprocal. In this case, the calculation of the channel gain in the downlink is the same calculation as the uplink. A similar configuration for optical wireless devices would facilitate the calculation process.

6.3. Link budget for communication with retroreflectors

6.3.1. *Principle of operation*

A retroreflector is a passive optic component that reflects the incidental light in the exactly opposite direction of this incident beam. The retroreflector generally has a large angle of vision (FOV). It is, for example, about ±12° for glass retroreflectors. The association of several retroreflectors makes it possible to increase the value of the field of vision.

The retroreflectors can also act as optical communication systems. If a retroreflector is associated with an electro-optical obturator on the cubic corner, it is possible to obtain a wireless optical communication. Indeed, the considered retro-beam can be activated or deactivated or at least modulated. By the assembly of a modulating retroreflector (MRR), we can allow the data transmission without the use of a light source such as laser or LED. For example, the basic station lights a module with a continuous or modulated optical beam. This beam arrives on the reflecting module and is passively considered toward the base station. The obturator, activated or deactivated by an electric signal, allows the transmission of the data coming from the module. This reflected signal containing the data from the module enters the receiver part of the base station to be treated. Figure 6.6 shows the diagram of an MRR.

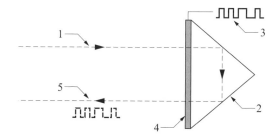

Figure 6.6. *Principle of retroreflexion*

The principle of retroreflexion is defined by an incidental beam (1) which is reflected on the cubic corner (2). A numerical signal (3) is treated by an electro-optical obturator (4) in order to transmit a modulated signal (5). The reflected beam is in the same direction as the incidental beam. The advantages of this system are that it is light, compact, and energy saving. Moreover, the retroreflexion is insensitive with possible micromovements to the base station and to its orientation.

6.3.2. *Optical budget*

The calculation of the optical budget described below is defined for a retroreflector of cubic corner type with an optical device called "cats eye" with which the reflectivity of the beam is most effective. Indeed, precise alignment is not necessary and such systems reflect the incidental beam independently of the exact orientation of the reflectors.

In such a wireless optical system, the optical beam is initially emitted from a base station A with a definite divergence in order to cover the zone in which is located the retroreflector B. This module receives part of the emitted beam, then reflects it under another observation angle. This reflected beam arrives then on the device of reception of the base station A.

In the literature, several authors examined the geometrical factors which affect the relation between the transmitted power and the received power [LEO 91, LAS 01, ACH 04].

The starting assumption is a complete coverage of the incident beam coming from the base station A on the module B.

An optical beam is emitted from an emission device A with a diameter of transmission d_t and a divergence angle Φ_t (radians). It illuminates completely a module B at a distance D (Figure 6.7).

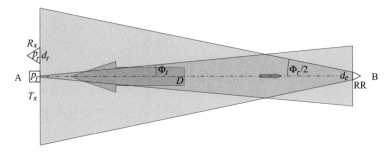

Figure 6.7. *Diagram of radiation of the emitted and retroreflected beam*

The reflected beam coming from the reflectors will have an effective diameter d_{re} and a divergence Φ_r (in radians) equivalent to the divergence of the transmitted beam and effects of diffraction due to the limited size of the reflectors. By supposing 100% of reflectivity of the reflectors and taking into account of a factor T corresponding to the atmospheric attenuation, the general formula of the received power P_r by device A on the effective surface of detection having a diameter d_r is given by:

$$
P_r = P_t \left(\frac{d_{re} d_r T}{(\phi_t D + d_t)\left(2 d_{re} + \dfrac{2.44 \lambda D}{d_{re}}\right)} \right)^2
$$

Equation 6.22. *Received power*

It is obvious that distance D is a significant parameter for the value of the received power. For systems of short distance, the atmospheric or molecular attenuation T is negligible and that:

$$2d_{re} \gg \frac{2.44\lambda D}{d_{re}}$$

Then by preserving the Φ_r term, equation [6.22] can be reduced to [ALH 06]:

$$P_r = P_t \left(\frac{d_{re}d_r}{(\phi_t D + d_t)(\phi_r D + d_r)} \right)^2$$

Equation 6.23. *Simplified received power*

As an example and with the values indicated in Table 6.2, we note that at a distance of 1 m, the received power is 0.01% compared to the emitted power. In the same way, when an optical source in A (Figure 6.7) emits 100 mW (20 dBm) with a divergence of ±3 mrad, the detection device in A will receive 0.01 mW (−20 dBm).

Parameters	Values
Distance – D (m)	3
Diameter of the receiver effective surfaces – d_r (mm)	2
Diameter of the retroreflector – d_{re} (mm)	8
Divergence of the reflected beam – Φ_r (mrad)	24
Divergence of the transmitted beam – Φ_t (mrad)	6
Diameter of the transmitted beam – d_t (mm)	2

Table 6.2. *Optical budget, typical values*

In this example, the losses due to the surface reflexions, glass absorptions, and interface losses are not taken into account. It indicates the best performance of the device.

6.4. Examples of optical budget and signal-to-noise ratio (SNR)

In the following examples, we propose to calculate the optical margin for several systems in various configurations. These simulations are carried out using the software

QOFI, adapted for wireless optical communications. It is a tool for simulation in three dimensions [TEC 11]. It makes it possible to simulate optical devices for wireless communications in confined space with calculation of the geometrical attenuation, the impulse response of the channel, and the coverage zone in a furnished room. This application makes it possible for the user to create an interior and a 3D model by dimensioning a room with the insertion of furniture, the base stations, and the modules. It proposes calculations in LOS, WLOS, and DIF while integrating:

– the room dimensions, reflexion of the room, the floor, and the ceiling;

– the object dimensions, their position and orientation with their reflexion properties.

This software is based on the Nettle's model, the technical details can be found in [BOU 08], and it is free downloadable via Internet sites [BER 08] or [QOF 11]. Table 6.3 presents several examples of wireless optical systems as shown in Figure 6.8. It supposes a power of emission of class 1, an ideal filter of transmission with a transfer function equal to 1, an ideal concentrator with a refraction index n of 1.52, and one OOK-NRZ modulation.

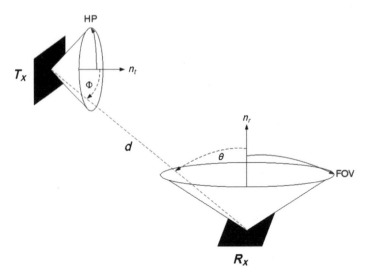

Figure 6.8. *Emitter–receptor link*

6.4.1. *Examples of optical budget*

Case 1 (WLOS): the bandwidth is 10 MHz, an OOK-NRZ modulation, a throughput of 10 Mbps, and the transmitter and the receiver are located on their respective normals (Φ and θ are null);

Case 2 (diffusion): the bandwidth is 10 MHz, an OOK-NRZ modulation, a throughput of 10 Mbps, and with retroreflectors (d_{re} = 10 mm, Φ_r = 30 mrad, and d_t = 4 mm);

Case 3 (WLOS): the bandwidth is 100 MHz, an OOK-NRZ modulation, a throughput of 100 Mbps, and the transmitter and the receiver are located on their respective normals;

Case 4 (WLOS with a part of diffusion): the bandwidth is 100 MHz, an OOK-NRZ modulation, a throughput of 100 Mbps in LOS, and a diffusion (wall and ceiling: painting, ground: parquet floor, T_x and R_x to 1 m of the wall and the ground), the transmitter and the receiver are located at angles different from their respective normals;

Case 5 (LOS): the bandwidth is 1,000 MHz, an OOK-NRZ modulation, a throughput of 1 Gbps, and the transmitter and the receiver are located on their respective normals.

	Case 1	Case 2	Case 3	Case 4	Case 5
B (MHz)	10	10	100	100	1,000
P_t (dBm)	20	20	20	20	20
HP (°)	±40	±30	±30	±30	±15
m	2.6	4.82	4.82	4.82	20
Angle Φ (°)	0	0	0	25	0
S_e (dBm)	−50	−50	−40	−40	−36
Photodiode $area - A_{ph}$ (mm^2)	78	78	7.1	7.1	0.3
FOV (°)	±30	±30	±20	±20	±15
Efficient area $- A_{eff}$ (mm^2)	726	726	140	140	10
Concentration gain (dB)	9.7	9.7	13	13	15
Angle θ (°)	0	0	0	15	0
Distance d (m)	9	7	7	4	4
Att_{geo} (dB)	53	69	55	54	56
P_r (dBm)	−32	−49	−35	−34	−36
M_l (dB)	18	1	5	6.14	0
Coverage to distance d (m^2)	179	51	51	51	3.6
d_{max} (m)	65	7	11	11	4
d_{limit} (m) with Φ =HP and θ = FOV	42		7.8		1.6

Table 6.3. *Five typical optical budget examples*

Figure 6.9 shows the characteristics of the impulse response in case 4.

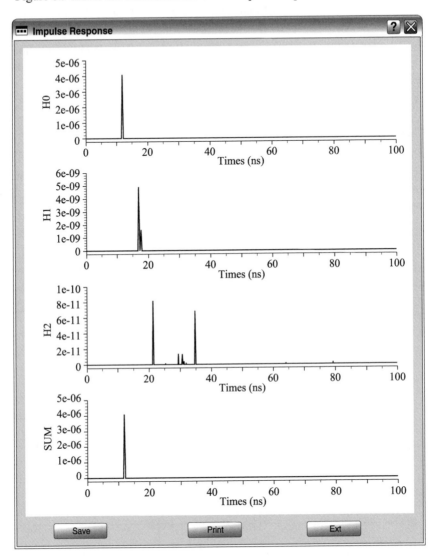

Figure 6.9. *Profile of the impulse response for case 4 (LOS + DIF)*

The contribution of the diffusion is the sum of the H1 (sum of the signals received with one reflexion) and H2 (sum of the signals received with two reflexions) curves. It accounts for only 0.01 dB in the margin link and created no intersymbol interference (ISI).

6.4.2. *Examples of SNR and BER*

From the expression of the SNR (signal-to-noise ratio, equation [4.12]), it is possible to present typical values of a wireless optical system in the case of an OOK-NRZ modulation:

$$SNR_{Optic} \cong \frac{R^2 H(0)^2 P_t^2}{2qRBP_b}$$

where

– R represents the proportionality factor of optic/electric conversion (or spectral sensitivity), for example $R = 0.6$ (A/W).

– $H(0)$ is the channel gain at the null frequency or the attenuation of the average optical power, for example $H(0) = 3.03.10^{-6}$.

– P_t is the transmitted average optical power, for example $P_t = 10$ mW (10 dBm).

– P_b is the noise power.

– $q = 1.6 \times 10^{-19}$ C$_b$, is the electron charge.

– B is the binary rate, for example $B = 150$ Mbps.

The binary error rate (BER) is a value relating to the error rate measured with the reception of a digital transmission. It is in correspondence with the attenuations or the level of disturbance of a signal transmitted in the channel. It is given by the following formula:

$$BER_{OOK}^{LOS/WLOS} = Q\sqrt{SNR_{woc}}$$

where

$$Q(x) = \frac{1}{\sqrt{2\pi}} \int\limits_{x}^{+\infty} e^{-y^2/2} dy$$

Equation 6.24. *Binary error rate*

We search the values of SNR and BER for various values of the power of the ambient light. Table 6.4 presents these values with a QoS indication.

P_{noise} (mw/dBm)	$10^{-4}/-40$	$3.2 \times 10^{-4}/-35$	$10^{-3}/-30$	$3.2 \times 10^{-3}/-25$	$0.01/-20$	$0.1/-10$
σ^2_{noise}	2.9×10^{-18}	9.1×10^{-18}	2.9×10^{-17}	9.1×10^{-17}	2.9×10^{-16}	2.9×10^{-15}
SNR (dB)	20.6	15.6	10.6	5.6	0.6	−9.4
BER	4.3×10^{-27}	8.4×10^{-10}	3.5×10^{-4}	2.8×10^{-2}	1.4×10^{-1}	3.7×10^{-1}
QoS	Operation			Degraded operation		

Table 6.4. *Typical values of power of ambient noise and SNR*

A significant value of the ambient optical noise can induce a degradation of the system. An increase in the optical assessment makes it possible to correct it. That is, translated, for example, with an increase in the sensitivity in reception, a reduction in divergence (DIV) in emission, or an increase in the emitted power.

Concerning the two parameters in emission, there is nevertheless a limitation of these values because of the regulation on safety with respect to the people. This element is in particular developed in Chapter 7.

Chapter 7

Immunity, Safety, Energy and Legislation

"Eyes are the windows of the soul"
George Rodenbach,
1855–1898

The emitted optical radiation of the emission module of an optical device is a very significant parameter in the calculation of the optical link budget. This radiation is characterized by the emitted power, its aperture (or field of view – FOV), and the emission wavelength. For reasons of safety, precautions or limitations can be considered according to the values of these three parameters. This is what will be developed in the first part of this chapter. The second part will approach the safety aspect of the communication, the third part evokes the power consumption, and, finally, the last part relates to the legislation.

7.1. Immunity

7.1.1. *International references*

Human beings, among others, can see wavelengths ranging from 400 nm (blue) to 700 nm (red), this is probably due to the evolution of the species, because this range corresponds to the peak of the solar spectrum (see Figure 3.12). Nevertheless, some tissues of the human eye can also interact with other ranges of the spectrum of optical radiation and are thus likely to be damaged under particular conditions.

In the near infrared (from 700 nm to 1,400 nm), radiations can be particularly dangerous for the retina of the eye (Figure 7.1), not only because of the process of concentration of the optical beam by the eye, but especially because of the absence of the palpebral reflex (blinking of the eyelid), for these non-visible radiations.

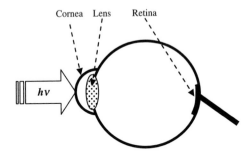

Figure 7.1. *Schematic cross section of an eye*

Beyond the near infrared (>1,400 nm), radiations are absorbed by the cornea of the eye and high optical powers can burn the cornea.

Under the visible light (<400 nm), i.e. in the field of the ultraviolet (UV), the effects of an excessive exposure on the skin are now well known (degeneration of cutaneous tissues or "photo-aging"). The acute exposure of the eye to UV can induce painful effects (conjunctivitis) which are reversible and it does not seem to create long-term lesions. Acute and chronic effects can appear as an appearance of a veil on the cornea or a cataract.

To eliminate any potential risk, there are international organizations which provide directives on safety for the transmission of a laser beam. Most significant is the International Electrotechnical Commission (IEC). IEC is an organization for standardization in the fields of electricity, electronics, and related techniques. The majority of its standards are developed jointly with the International Organization for Standardization (ISO). IEC is a standards' organization dealing with areas of electricity, electronics, and related technologies. IEC was set up in 1906 and is currently based in Geneva. More than 60 countries are now members of IEC.

In the field of the radiation from a laser beam, IEC published several standards, the principal ones in connection with telecommunications are:

– IEC 60825-1 (2008): Safety of laser products – Part 1: equipment classification and requirements;

– IEC 60825-2 (2008): Safety of laser products – Part 2: safety of optical fiber communication systems (QFCS);

– IEC 60825-12 (2004): Safety of laser products – Part 12: safety of free-space optical communication systems used for transmission of information.

In the field of light-emitting diodes (LED) functioning in the visible spectra, the European Commission recommends Directive 2006/25/CE on the protection of workers against exposure to risks due to the artificial optical radiations. There are two well-known risks:

The adverse effects due to sparking: The variation of the radiation power of a visible light can induce an undesirable physiological reaction like headaches or a loss of concentration [KUL 98]. In extreme cases, it can cause epileptic fits in a minority of the population [HAR 94]. However, the frequencies of modulation for the majority of the wireless optical communication systems are higher than 500 kHz and are thus much higher than the highest limit of the cutoff frequencies for the adverse effects mentioned.

The temporary blindness: Because of the strong luminosity of some LEDs, it is possible to undergo a temporary blindness with an effect of retinal remanence [REI 09]. The blindness effect, caused by an overload of the retinal receivers, can persist up to 5 s; whereas the effect of retinal remanence can persist up to 20 s, even when the powerful light resulting from the LED is seen from several meters.

Standard IEC 60825-1 (last edition 2008) and its European integral version IN 60825-1 [EN608] are regarded as the reference for wireless optical devices of data transmission in confined space using a coherent beam.

For the measurement of a safety level of a laser source, this standard defines a parameter known as the maximum permissible exposure (MPE); it is the electromagnetic radiation level to which a human observer can be exposed during a certain time without incurring immediately or later, harmful effects.

The second parameter is the accessible emission limit (AEL). This value is derived from MPE and allows class definition of laser equipment. The classes categorize a level of risk or harmlessness. It is obligatory and references mention all apparatus containing a laser source.

7.1.2. *Type of laser classes*

The standard IN 60825-1 indexes lasers in seven classes, numbered from 1 to 4, and whose risk increases from class 1 to class 4 (Table 7.1). The limit of each class is based on the power or the energy values emitted by the laser sources and accessibility to the user.

Class	Fields
1	A class 1 laser is safe under all conditions of normal use. This means that the MPE cannot be exceeded.
1M	A class 1M laser is safe for all conditions of use except when passed through magnifying optics, such as microscopes and telescopes. Class 1M lasers produce large-diameter beams or divergent beams. The MPE cannot normally be exceeded unless focusing or imaging optics is used to narrow the beam. A laser can be classified as class 1M if the total output power is below class 3B but the power that can pass through the pupil of the eye is within class 1.
2	A class 2 laser is safe because the blink reflex will limit the exposure to no more than 0.25 s. It only applies to visible-light lasers (400–700 nm). Class 2 lasers are limited to 1 mW continuous wave, or more if the emission time is less than 0.25 s or if the light is not spatially coherent. Many laser pointers are class 2.
2M	A class 2M laser is safe because of the blink reflex if not viewed through optical instruments. As with class 1M, this applies to laser beams with a large diameter or large divergence, for which the amount of light passing through the pupil cannot exceed the limits for class 2.
3R	A class 3R laser is considered safe if handled carefully, with restricted beam viewing. With a class 3R laser, the MPE can be exceeded, but with a low risk of injury. Visible continuous lasers are limited to 5 mW.
3B	A Class 3B laser is hazardous if the eye is exposed directly, but diffuse reflections such as from paper or other surfaces are not harmful. Continuous lasers in the range from 315 nm to far infrared are limited to 0.5 W. For pulsed lasers from 400 nm to 700 nm, the limit is 30 mW. Other limits apply to other wavelengths and to ultrashort lasers.
4	Class 4 lasers include all lasers with beam power greater than class 3B. By definition, a class 4 laser can burn the skin, or can cause potentially devastating and permanent eye damage as a result of direct or diffuse beam viewing. These lasers may ignite combustible materials, and thus may represent a fire risk. Some industrial, scientific, military, and medical lasers are in this category.

Table 7.1. *Classification of the lasers*

According to the laser classification of the equipment, protection and safety devices are possibly implemented (cover of protection, operation with key, blocking of access, etc.). Complementary, the standard indicates the obligatory or optional informations to show on the equipment. This informations will be different according to the laser class and are indicated in Table 7.2.

Class	Fields
1	Each laser of class 1 must carry an explanatory label being marked "Laser class 1". If not, to replace the label and with the discretion of the manufacturer, the same declarations can be included in the guide of information for the user.
1M	Each laser of class 1M must carry an explanatory label being marked "Radiation laser. Not to look at directly with optical instruments. Laser class: 1M". If not, to replace the label and with the discretion of the manufacturer, the same declarations can be included in the guide of information for the user. It is possible to add after "instruments" the mention "binocular or telescopes". If the AEL exceeds the value of the class 3B, the mention "the exposure of the skin close to the opening can cause burns" must be added.
2	Each laser of class 2 must carry an explanatory label being marked "Radiation laser, not to look at the beam, laser. Laser class 2".
2M	Each laser of class 2M must carry an explanatory label being marked "Radiation laser, not to look at the beam or to see directly with optical instruments, laser. Laser class 2M". It is possible to add after "instruments" the mention "binocular or telescopes". If the AEL exceeds the value of the class 3B, the mention "the exposure of the skin close to the opening can cause burns" must be added.
3R	Protective eyewear is typically required where direct viewing of a class 3B laser beam may occur. Each laser of class 3R must carry at least an explanatory label being marked "Radiation laser. Avoid the direct exposure of the eye, laser. Laser class 3R". It is possible to replace the mention "Avoid the direct exposure of the eye" by "Avoid the exposure to the rays".
3B	Class 3B lasers must be equipped with a key switch and a safety interlock. Each laser of class 3B must carry at least an explanatory label being marked "Radiation laser. Avoid the exposure to the rays. Laser class 3B".
4	Class 4 lasers must be equipped with a key switch and a safety interlock. Each laser of class 4 must carry at least an explanatory label being marked "Radiation laser. Avoid the exposure of the eyes or of the skin to the direct or diffuse radiation, Laser class 4".

Table 7.2. *Legal mention of information*

In light of information coming from Tables 7.1 and 7.2, a wireless optical equipment of numerical data transmission for general public use must be of class 1. Indeed, a laser of class 1 is not dangerous even if the beam is visualized with an optical instrument such as a telescope or a magnifying glass.

Standard 60825-1 more precisely characterizes the apparatuses of class 1 in the following way (Appendix C of the standard): "laser apparatuses which are without danger during the use, including the visualization of the direct beam in the long run, even when the exposure occurs with instruments of optical visualization (magnifying glasses or binoculars). Class 1 also includes lasers of strong power which are entirely locked up so that no potentially dangerous radiation is accessible in the course of use (embarked laser products)".

A remark is added for the beams emitted in the visible spectrum from a laser: "the visualization of the beam of a laser product of class 1, which emits in the visible one, can still produce dazzling visual effects, in particular with a weak ambient light".

7.1.3. Method for calculation

The standard provides a table with which AEL can be calculated for class 1 according to various factors, such as the wavelength, the duration, and the size of the image on the retina of the eye. The size of the image on the retina is the surface of the optical beam which arrives at the retina after concentration of the beam. The result of this focusing on the retina will be different if the laser source is punctual (low size with a low divergence of the beam) or wide (strong divergence of the beam, e.g. a lamp) as shown in Figure 7.2.

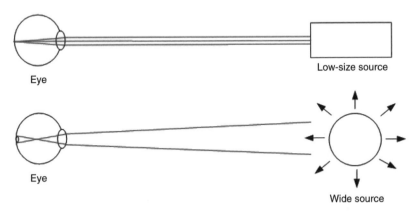

Figure 7.2. *Low size and wide sources*

In addition, there are three conditions of measurement specifying the aperture diameter and the distance from the source in order to determine the laser class of this source.

Condition 1: intended for a collimated beam (parallel rays) where telescopes and binoculars can increase the danger;

Condition 2: intended for the divergent sources, where the use of magnifying glasses or microscopes can increase the danger;

Condition 3: apply to the naked eye.

In the standard, it is explicitly mentioned that the condition of the most restrictive measurement must always be applied, i.e. the power accessible from a laser of class 1, measured at the exit of the measurement aperture, should never exceed AEL from class 1 for these three conditions.

Moreover, two sets of evaluation are specified in correspondence with each of the three measurement conditions:

– A simplified evaluation presentation (by default) that covers all the point sources. The three conditions are associated with the measurement process for various values of distance r and aperture diameter d (Figure 7.3); the power emitted at the measurement aperture should not exceed AEL; the simplified evaluation can be applied to wide sources and that induces more restrictive limits.

– For the wide sources (e.g. a laser with an optical diffuser), a second more complex evaluation method can be applied, with less constraining emission limits. When this method is applied, a correction factor is included in the process of research of AEL value.

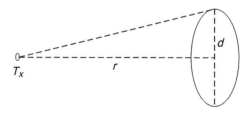

Figure 7.3. *Configuration of measurement*

Several factors of correction are introduced into the formula of power of accessible emission limit (PAEL). They depend on the wavelength, the diameter, the measurement aperture, and the duration.

Indeed, AEL at the output of the measurement aperture depends also on the duration. The standard prescribes a duration of 100 s for the class 1 products if the intentional vision in the long term is not inherent in the design or the function of the product. This duration must be taken into account in the range from 400 nm to 700 nm (visible range) but in infrared field (higher than 700 nm); the characteristics of calculation remain unchanged for durations higher than 100 s.

Figure 7.4 shows, in the form of graph, several results of the computing process in the field of infrared [WOL 08]. In this spectral range, the power of the PAEL is similar to the power of the maximum permissible exposure (PMPE). These results are presented in terms of maximum acceptable radiation intensity (source of emission in one solid angle unit) according to the wavelength, for various diameters (d) of the emission source. The curve $d = 0$ corresponds to a point source.

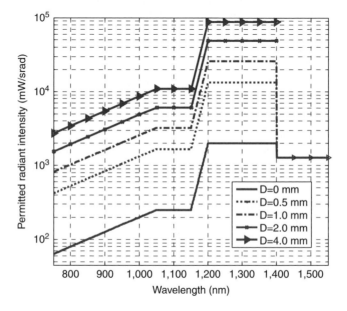

Figure 7.4. *Allowed radiation intensity class 1 [WOL 08]*

The results were calculated and presented with the most restrictive condition:

– for the punctual sources, condition 2 with the default method ($d = 0$);

– for the wide sources, condition 3 with the more complex evaluation method;

– for remind, the relation between the normal intensity radiated in the normal direction $I(0)$ and the power is (equation [6.4]) given by:

$$I(0) = \frac{m+1}{2\pi} P_t$$

And the steradian defines the solid angle. It is calculated as being the solid angle that, having its top in the center of a sphere, cuts, on the surface of this sphere, a surface, most often circular, equivalent to that of a square whose side is equal to the ray of the sphere. A steradian (Sr or srad) of solid angle corresponds to a plane angle whose top is approximately 60° or ±30°.

Consider the Cases 2 and 5 of Table 6.3 (Chapter 6). The power values are 20 dBm (100 mW) with an aperture angle of ±30° ($m = 4.82$) and ±15° ($m = 20$), respectively. The values of the normal intensity $I(0)$ radiated in the normal direction will thus be 92.6 mW and 334.2 mW. If the proposed system functions with a wavelength of 850 nm and with an aperture diameter of 1 mm, the graph indicates us a maximum acceptable radiation intensity lower than 1,200 mW. Therefore, the two products, from Case 2 and 5, are indexed in class 1.

As well as the eye, the human skin is also sensitive to damage caused by optical radiations. The standard also specifies an MPE for the skin, which must be measured at the point where the exposure level is highest. This value also depends on the wavelength, the diameter of the aperture window, and the distance between the emission source and the directly accessible point. For example, in the most critical case, if the source is directly placed on the emission window with a beam diameter lower than 3.5 mm, the value of the maximum power MPE would be 100 mW.

The implementation of the emission devices of a wireless optical system requiring a minimal focal distance to make the beam divergent makes improbable this configuration.

Nevertheless, the emitted total power must be lower than 500 mW to be in conformity with class 1 for the requirements on skin safety [WOL 08].

7.2. The confidentiality of communication

The confidentiality of a communication between two interlocutors is assured if a third person or an unspecified detector cannot collect exchanged information. To ensure this safety, there are physical solutions by material techniques and numerical solutions by information coding, the last field is called cryptography.

7.2.1. *Physical confidentiality*

The fundamental element is that the light of the optical spectrum (ultraviolet, visible, and infrared) does not pass through walls. The optical transmissions are thus confined inside a room and they are consequently naturally better protected.

In spite of the existence of an aperture in a room, such as doors and windows, the visible signal becomes so attenuated and diverging that the possibilities to collect part of the beam become difficult. Hacking of information is more restricted than for the radiocommunications.

Moreover, the insertion of blinds or curtains window level brings a notoriously effective protection. Moreover, the existence of materials placed on windows, with different coefficients of transmission according to the wavelengths, can offer an additional safety [BOU 04]. Concerning the possibilities of propagation in a corridor or another part, the closing of a door offers, this, level of a simple and effective safety.

Finally, the non-deterministic wavelength jump solution presents one possibility of physical safety (related to a synchronization algorithm) of communication up to that point unequalled.

7.2.2. Numerical solution

The technique used is called the cryptography and several possibilities of cryptography are offered, from classic key cryptography to quantum cryptography [SCH 96, SIN 01]. Confidentiality of a communication between two communicators is assured if a third person or an unspecified detector cannot collect exchanged information. To do that, we can either code information, this is the domain of cryptography, or prohibit (by material techniques) beams of information to go to places where it is not wished.

7.2.2.1. Cryptography

The confidentiality of the communications can be of strategic importance as much for economic questions as for military ones; it is also the protection of privacy. To ensure confidentiality, the interlocutors must then return the contents of their messages, incomprehensible to a third person or more likely to one possible spy. The technique used is the cryptography, which literally is the art of to hide (crypto) a written text (graphy). Whatever the technique used in cryptography, the guiding principle is always the same. A transmitter, called traditionally Alice, codes her message using a key. The result is an encrypted message, which she sends to her correspondent called traditionally Bob.

This Person-Bob, using key deciphers receives the message and can read it. Confidentiality correspondence between Alice and Bob is guaranteed owing to the fact that a third person, who is also traditionally called Eve as a spy, does not have the deciphering key. It is permitted that the safety of an encryption system does not depend on the safeguarding of the secrecy of the algorithm, but only of the key secrecy.

Since the beginning of cryptography, we have seen an escalation in activity between cryptographs and code breakers. When the codes do not resist to Eve any more, Alice and Bob must change their cryptographic protocol. In addition, protocols must be increasingly complex so that it is necessary to change them as little as possible.

7.2.2.2. *Public and secret key cryptography*

In electronic communication, we are currently using RSA (Rivest, Shamir, Adleman) code which is a public-key code, thus avoiding the constraint that the two interlocutors must obligatorily first exchange the secret key. The safety of the algorithm rests on a mathematical conjecture, which stipulates that it is very difficult to calculate p and q prime numbers from their product. This conjecture was established from the mathematical properties of the best known factorization algorithms and the computing power of the computers. Nevertheless, it is not proved that the current factorization algorithms are optimal. As for the computers, power doubles, according to the Moore law, every 18 months. Increasingly large keys are factorized. Currently, we regard as sure those 1,024 bits keys at least. This key could be factorized in 2,048 bits, taking into account the known evolutions.

Based on unproven mathematical conjectures, although simple to implement since there is no exchange of secret key, RSA algorithm does not allow absolute communication safety. For safer exchanges, we can turn toward secret key codes. There is then no complex algorithm to use. But the key must be as long as the coded message and should be used only once. These conditions mean that this type of cryptography is used only at the diplomatic and military level.

Therefore, we have on the one side a public-key algorithm which does not require the exchange of a secret key but whose absolute safety is not guaranteed. And on the other side, we have an infallible proven algorithm, which requires the exchange of a secret key at the time of each mail.

7.2.2.3. *Quantum cryptography*

The only coding technique to guarantee the perfect inviolability of the communications, quantum cryptography rests on the exchange of single photons between the two correspondents. Quantum cryptography takes advantage of the Heisenberg uncertainty principle. Some characteristics of a quantum state cannot be measured simultaneously: whoever measures one modifies the others. The quantum state represents the key here allowing us to decode the message.

We can code a bit of information on the state of polarization of a single photon (qubit), but this information could be read only if we know the base of polarization in which it was coded.

The property of the quantum mechanics which will guarantee the safety of transmission of the key is the not-cloning theorem. It is not possible to copy an unknown quantum state perfectly: a spy cannot copy a qubit. We can thus distribute a coding quantum key (QKD – quantum key distribution), by coding information on the state of polarization of a single photon. When the key is as long as the message, the confidentiality is total, any mathematical attack is able to break it.

The luminous impulses are single photons, sent one-by-one, and Alice knows their state of polarization. To handle these single photons, it is necessary to have sources of single photons and detectors of such single photons.

7.2.2.4. *Quantum telecommunications in free space*

Quantum transmissions in free space were carried out, but more often starting from approximations of single photons sources.

From an attenuated coherent source: this method, generally applied to simulate a source of single photons, consists of strongly attenuating a photon beam from an impulse laser. For an average number of 0.1 photon by impulse, there are 90.5% of empty impulses, 9% contain a photon, and 0.5% contain two photons. These statistics of the number of variable photons according to the impulse are not very satisfactory for true communications although it is interesting for laboratory experiments.

The first experimentation in 1992 was made on a distance of 30 cm in free space [BEN 92]. The known record by the authors is an exchange of secret key in free space on a distance of 23 km in mountain environment [KUR 02]. The climatic fluctuations and parasitic lights are then the essential problems.

Single nitrogen-vacancy centers (called NV centers) in a diamond nanocrystal emit photons with a wavelength around 637 nm [BEV 02]; by dilution of these NV centers in the matrix, the authors carried out a single photons source. On a distance of 50 m in free space, approximately 8,000 secret bits per second were transmitted. This system allows a quantum key distribution "on request", but not at wavelengths suitable for free-space telecommunications.

Another single photons source is proposed starting from erbium ions diluted in a matrix-host with large gap, a dielectric such as silica, for example, [DEF 01] and which the emission of the single photon at 1.55 μm emitted by only one exited erbium is collected in a near field by a thinned fiber before being re-emitted in free space. This technique allows us to have really a source of single photons at the traditional wavelength of optical telecommunications (1.55 µm) and at the ambient temperature. Due to the flexibility of optical fiber, here we have a very great simplicity of implementation that must facilitate the alignment problems between the transmitter and the detector. For optical links in free space requiring an absolute secrecy, this type of quantum keys exchange can prove absolutely useful.

7.2.2.5. *Non-encrypted connections in confined space*

The confidentiality guarantee of the non-encrypted communications in confined spaces imposes to limit the maximum transfer of information out of the limits of this space where the communications take place. In a meeting room, researchers [DEF 02] showed that the use of structured artificial materials can guarantee the confidentiality. Structured artificial materials with two or three dimensions would allow us to have perfect reflectors.

7.3. Energy

The search for communication systems with low power consumption is more and more sought after and this parameter now influences the behavior of the consumers. In this respect, the energy comparison is one increasingly significant and relevant aspect in the choice of a technology.

It is generally difficult to compare the energy consumption of various systems. The total consumed energy is constituted from the energy necessary to transmit the data as well as the power consumption for the signal and data processing. Table 7.3 presents an example of comparison and indicates that optical wireless communications can present a more economic energy/debit ratio. Because of this least power consumption and of its compactness, the wireless optical system is more and more used in the satellites [SOD 09].

Standard	Power (W)	Data rate (Mbps)	Used energy (J/Mb)
WiFi – IEEE802.11(g) [DLI 05]	3	54	5.5×10^{-2}
Wireless optical equipment [OME 10a]	10	1,250	0.8×10^{-2}

Table 7.3. *Energy comparison*

Moreover, because of the characteristic of the transmission media, i.e. the use of an optical beam in visible or an infrared field, it is possible to re-use, in addition to daylight, the optical energy sent for the data communication.

For example, let us take a smart object in wireless optics with a consumption of 10 W which functions 10 h/day. The power consumption per day of this equipment is thus 100 Wh.

The electric production of a photovoltaic plate of dimension A4 (0.06 m^2), without orientation or tilting optimization in a room, can reach approximately 20 Wh/day [ENE 11]. It is thus currently possible to obtain, with the example of consumption and in the current state of the photovoltaic output, more than 20% of free energy per day.

7.4. Legislation

7.4.1. *Organization of regulation activities*

International Telecommunication Union (ITU), whose headquarter is in Geneva (Switzerland), is an international organization under the United Nations system within which States and private companies coordinate global telecommunications services and networks. The ITU is structured in three distinct sectors:

– ITU-D: Development;

– ITU-T: Standardization;

– ITU-R: Radiocommunication.

All work concerning radiocommunications is concentrated in the last sector and manages in particular the uses of the radio-frequency spectrum. This revision of attributions of the plans of frequency and the division of the spectrum is carried out during the World Radiocommunication Conference (WRC) which meet at the end of each operating cycle of ITU-R study group. A WRC meets approximately every 3 years to develop, adopt, and revise the radio regulations (RR). The WRC decides attribution of the frequency bands to various services of radiocommunications (fixed service, mobile, broadcasting, satellite, radiolocation, space research, exploration of the Earth, radioastronomy, etc.). Each ITU country must conform to these attributions and these conditions of sharing are fixed by ITU. Attributions, access, and sharing conditions are described in the RR. For information, the next WRC will be held at the beginning of 2012, in Geneva (Swiss).

At the European level, the organization of reference is the European Conference of the Posts and Telecommunication (CEPT). CEPT groups 44 countries, including the 27 members which are from the European Union. The permanent office, the European Radiocommunications Office (ERO), is based in Copenhagen (Denmark). By the decisions and recommendations of the Committee of the Electronic Communications (the CEC), CEPT decides, within the framework of the attributions fixed by ITU-R, the particular conditions which prevail for the use of these attributions: reservation of frequency bands for particular systems (GSM, DECT, etc.), determination of channels for the fixed service in various frequency bands which are allocated to it by ITU-R and technical and lawful conditions of operating. These decisions or recommendations aim to harmonize the frequencies used by the various European countries, with the objective to facilitate the development of a European market, as well as deal with the problems of coordination at the borders. The various European countries must conform to the decisions of CEPT while the recommendations aim to harmonize the use of the spectrum without having a constraining character.

The acceleration of the rhythm of the WRC and the need for arriving more easily to a consensus within the framework of these conferences led the European countries to gather their forces together carry out the preparation of the WRC. The role played by CEPT within this framework is fundamental, the European Commission being until now the only witness. The Conference Preparatory Group (CPG), a working group of the CEC, coordinates this effort and prepares the common European positions to the WRC.

7.4.2. Regulation of wireless optical equipment

If we take again Article 1.5 of the RR, "1.5: radio waves or Hertzian waves: electromagnetic waves whose frequency is by convention lower than 3,000 GHz, being propagated in space without artificial guide", it appears that the wavelengths used by the wireless optical equipment are not currently covered by the clauses of the RR that are limited to the frequencies lower than 3,000 GHz. In fact, the wireless optical equipment operates at frequencies located between 150 THz and 500 THz.

This is why, there exists, at the present time, no legislation or management and attribution of this part of spectrum. One of the direct consequences of this characteristic is the absence of tax or expenses related to an attribution license.

It should, however, be noted in 2002, that the ITU Plenipotentiary Conference, which is the supreme body of the ITU, noting well that techniques of radiocommunication showed that it was possible to use electromagnetic waves in space, without artificial guidance at frequencies higher than 3,000 GHz, adopted a new resolution on the study of the use of the spectrum above 3,000 GHz (Resolution 118 – Marrakech, 2002). Consequently and during the WRC 2007, a resolution (Resolution 955 – Geneva, 2007) proposed to consider a process of installation of free-space optics (FSO), with definition of spectral limitation and measurement, in order to allow the sharing with other services, if allocation to various services in the RR above 3,000 GHz is considered feasible. This proposition is discussed during the WRC 2012.

Chapter 8

Optics and Optronics

8.1. Overview

We discuss the "physical layer" of wireless optical terminals in this chapter. The digital communication part and the communication protocol part will be discussed in Chapter 10 (OSI layers 1 and 2 and the TCP/IP layer 1).

The other layers are application oriented and will not be considered in this book.

The different modules of a wireless optical terminal are illustrated in Figures 8.1 and 8.2.

8.2. Optronics: transmitters and receivers

8.2.1. *Overviews on materials and structures*

The transmitter and receiver components of this section are components of very small dimensions (volumes below 1 mm^3) processed by deposition of thin films on semiconductor materials.

A crystal of semiconductor material is a poor conductor. The electrons which are the charge carriers are not free to move within the material when they occupy energy states in the low-energy valence band. They can only contribute to the current flow when they get enough energy in excess of the bandgap and are free in the conduction band as shown in Figure 8.3.

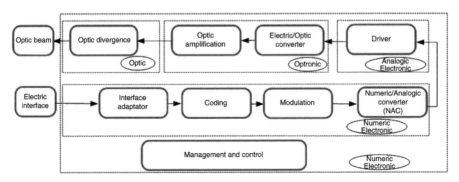

Figure 8.1. *Diagram of an optical wireless transmitter*

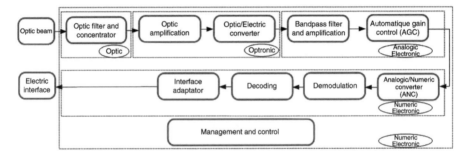

Figure 8.2. *Diagram of an optical wireless receiver*

This excess in energy needed to gain kinetic energy and be admitted into the conduction band may be the result of several physical phenomena, such as:

– an increase in temperature of a few K generally considered as a "parasitic phenomenon" (leakage current drifts depend on temperature variations, etc.);

– the presence of light quanta, photons, whose energies depend on their colors, in other words their frequencies, the wavelengths in vacuum, and the values of their energies compared to the bandgap energy.

It must be noted that when an electron from the valence band is well accepted in the conduction band, it releases its original condition and yields a positive hole charge carrier, which is also free to move. Thus far when the energy of an incident photon is greater than the energy gap E_g, it gives rise to an electron–hole pair when the photon is absorbed inside the semiconductor material.

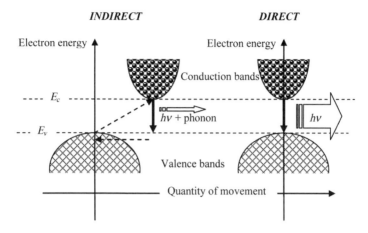

Figure 8.3. *Band diagram of a crystal of semiconductor material*

The condition on the frequency and wavelength is given by (equation [8.1]):

$$h\upsilon > E_g \text{ so } \lambda < \lambda_c = \frac{hc}{E_g} \text{ cutoff wavelength}$$

Equation 8.1. *Cutoff wavelength*

Next to the notion of electron–hole pair, we must introduce the notion of lifetime. An electron of the conduction band may be solicited to lose part of its energy, or take a state occupied by a hole in the valence band. The electron loses then a quantum of energy, may either find itself trapped on a level of impurity or defect in the crystal, or recombines with a hole when losing a quantum of energy at least equal to E_g.

In some semiconductor materials to which we will soon come back, the loss of this quantum of energy can result in the emission of a photon whose wavelength in vacuum is given by the following relations:

$$\lambda_g = \frac{h.c}{E_g} = \frac{1.24}{E_g} \text{ ; } \lambda_g \text{ is in } \mu\text{m and } E_g \text{ in e.V.}$$

Equation 8.2. *Emission wavelength*

The crystals of semiconductor materials are assemblies of one or more different atoms according to geometric rules in space which are also different. The physical properties and in particular the optical and electrical properties can be very different between materials. For example, looking at columns III and V of the Mendeleev's table, when we compare silicon (Si) and arsenic (As), gallium (Ga), phosphorus (P), indium (In), and aluminum (Al) compounds, we discover that the probabilities of radiative electron transitions between conduction band and valence band are very different. These transitions "conduction band valley-to-valence band summit" are indirect in the case of silicon and consequently unlikely, compared to the direct transitions, and therefore very likely in the case of the semiconductor "III–V" compounds. Energy gaps and valence–conduction band shapes of the III–V compounds lead to very different spectral properties, the spectrum of visible to near- infrared spectral range of fiber optic transmission telecommunications.

The semiconductor sources of the light waves detected by silicon photodiodes in the visible to near-infrared spectral range (0.4–1 μm wavelength) are widespread in the market of optoelectronic components and devices. The industrial development of sources and detectors operating in the infrared spectral range from "1 to 1.6 microns" for "Fibre To The Home (FTTH)" applications is still undergoing a high growth rate.

8.2.2. *Light sources*

Most of the sources made from semiconductor materials have the following advantages:

– their size, as we have previously mentioned a few mm^3 deposited on the bases of a few mm²;

– the near field of the exit pupil and the radiation pattern: well-known, stable, and reproducible;

– the possibility for modulating the flow of energy and the optical frequency power spectrum directly by the injection current: we can reach several tens of gigahertz, frequency dependent of the internal structure of the component.

Their structures are junctions elaborated by deposition of thin layers of III–V compounds (a few hundred nanometers to several microns thick) on a GaAs or InP. Radiative transitions and spontaneous emission occur in the junction area of these diodes when a forward injection current is applied. The spontaneous emission spectra depend on the III–V compound band structures and the gap energy which depend themselves on the chemical compositions of the compounds.

We will see later on in this section the different operating modes and optical behavior of these light sources.

8.2.2.1. *Light-emitting diodes (LEDs) and spontaneous emission*

Electrons and holes injected into the junction will fill the energy levels in the bottom of the conduction band and the top of the valence band, respectively. These two types of carriers will be required to recombine by emitting photons whose energies are quite different. Therefore, the spectral width of the emission will be quite large.

The "output radiant power vs injected current" curve is approximately linear up to a few milliwatts per square mm, limited by the heat dissipation of the junction.

Two junction structures lead to two types of exit pupil and radiation pattern:

The top surface-emitting diodes (Figure 8.4(a)): There are no plans to impose the directions of propagation of photons emitted inside the "chip"; the diameter of the pupil is the same as the opening of the upper contact surface; the radiation pattern is said to be Lambert's law (sphere of revolution around the z-axis), sphere tangent to the surface (Figure 8.5(a)).

The edge-emitting diodes (Figure 8.4(b)): The current is injected through an electrode – ribbon through a light guides between the two sides of the emissive "chip" (Figure 8.5(b) – the index n in the (x, z) is different from that in the (y, z)).

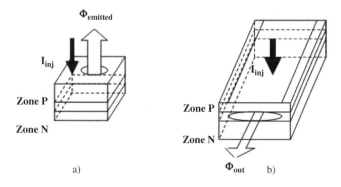

Figure 8.4. *LED spontaneous emission structures*

These two sets of structures from industrially mature technologies are very reliable. Both are designed and built to transmit near-infrared waves and are used in many electronic and optical applications (remote controls, emission in the optical fiber links, etc.). Costs are moderate due to the volume of the market.

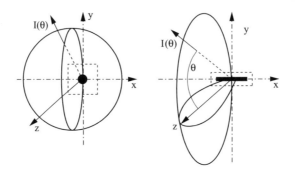

Figure 8.5. *Lambert's law: (a) I(θ) = I(0).cosθ and (b) I(θ) = I(0).(cosθ)n*

The following section is entirely devoted to surface emitting white light LEDs. The optical energy is distributed in the visible region from 380 nm to 780 nm. Their first applications are lighting and signalization.

8.2.2.2. White LEDs or visible light communication (VLC) LED

New LED structures, usually in SMC configuration (surface mounting components), have emerged since the 1990s. They aim to replace, over time, fluorescent or incandescent light sources. These new generations of LEDs provide a light output greater than that of an incandescent lamp and a performance/cost ratio (ability of light and life) that approximates more and more preferably the fluorescent tubes. In addition, technical progress in order to obtain ever more powerful LEDs, with warmer colors, and higher yields.

White LEDs are devices currently planned for lighting, but it is possible to modulate the intensity with relatively short switching times. The switching time is the time required to grow from a current of 10% to 90% of its final value or decreases from 90% to 10% and it depends on the capacity of the junction in the diode. This feature allows data rates relatively high in digital data transmission by optical fibers, for example. They have, in this regard, a number of advantages, including being powered by the industry and we can imagine a coupling between data transmission over power line and wireless optical transmission.

There are two main types of white LEDs:

– The first one is made of three LED colors: red, green, and blue (Figure 8.6). This light source is called multi-chip white LED or RGB LED. By modulating the intensity of one or more color, it is possible to achieve an optical digital transmission with higher data rates. However, the simultaneous operation of these LEDs can be tricky and the aging of each different color can degrade the quality of white light.

– The second approach is more recent. A layer of fluorescent material is deposited over a blue LED. This layer emits yellow light when excited by the blue light LED (Figure 8.7). The mixture of blue and yellow light creates an almost white light. For data communication, the blue wavelength is used because of its short switching time. To avoid the bluish appearance, changes are implemented, such as blue LED combined with a multi-phosphors or UV (ultraviolet) LED with phosphors RGB (red, green, and blue).

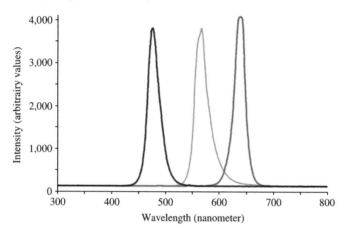

Figure 8.6. *Multi-chip LED spectra*

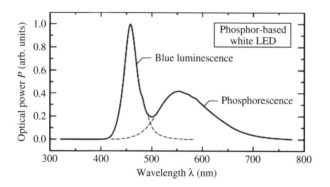

Figure 8.7. *Phosphor-coated LED spectra*

8.2.2.3. *The semiconductor laser structure*

The first amplification by stimulated emission structures developed on substrates of semiconductor materials was proposed by Professor Aigrain in 1958 and completed in 1962. Based on the calculations of Albert Einstein, the population inversion, which contradicts the thermodynamic equilibrium between energy levels and leads to the phenomenon of amplification, can be achieved by a strong injection of carriers in the conduction and valence bands.

Going back to the structure of LED emitting "edge-on". They consist of a heterojunction – epitaxial thin films of III–V materials with different compositions – which comes in the form of a planar dielectric waveguide with rectangular section. The carriers injected by the contact strip recombine radiatively in the contact and within the guide along which the photons created can spread. When the injected current density is high enough, there is inversion of the populations and the effect of gain occurs as it can be seen on the output radiant power versus the injected current curve. The diode becomes super-luminescent, high brightness, and the spectrum of the transmitted wave is narrower.

In the case of "Fabry–Perot" lasers (Figure 8.8), the waveguide is limited by two reflective facets cleaved along two parallel crystal planes and operates as a resonant cavity. In the case of single-mode distributed feedback lasers, the resonance can be obtained along a Bragg grating (periodic corrugation along the waveguide axis), a grating whose function is to select a single spatiotemporal mode.

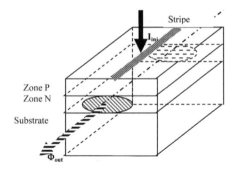

Figure 8.8. *Laser structure: thin epitaxial layers, stripe and Fabry–Perot resonator*

When the injection current I is sufficiently high and exceeds the threshold current, the gain is almost equal to the total losses of the cavity which are the propagation losses of the guide (diffusion, absorption, defects at the interfaces between epitaxial thin films, etc.) and losses at the level of reflective facets of the resonator.

Thus, in general, the active portion of semiconductor lasers has a dielectric waveguide structure buried in thin layers of materials of different compositions and electronic and optical properties. Unlike the LED spontaneous emission, the laser is an amplifier/generator of monochromatic light waves (low spectral width), consistent (the photons are in phase), with highly directional exit pupils from a few microns square. They are structurally more complex than emission LED with a level of modulation speed, power, and higher performance, but also a higher cost.

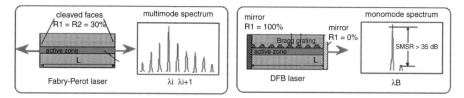

Figure 8.9. *Fabry–Perot and DFB lasers*

There is a second type of laser structure called vertical external cavity surface emitting laser (VCSEL) – a laser emission from the surface with a resonant cavity and amplifier composed of vertical Bragg mirrors. The VCSEL is cheaper than the structures emitting laser facets cleaved. Its radiation pattern is of a revolution, the circular exit pupil, its power consumption, and output radiant power are all lower.

8.2.2.4. *Synthesis*

When we look for the optimal solution, we must take into account the following: either high transmission power, fast modulation, divergence large enough to cover an important area or more transmitting device and security Class 1.

The electrical/optical converter is usually the association of a laser power driver and an automatic transmit power control. This device performs an operation against amplification loop feedback for controlling the signal to be transmitted with sufficient power to allow a correct reception.

The main element of the transmission system is the optical source. Most systems use laser diodes (LDs) due to their speed of response and radiant power output. The optical power can be increased by inserting an optical fiber amplifier-type erbium-doped fiber amplifiers (EDFAs), but the cost and congestion currently limit this option. The infrared LED light sources are also considerations in the implementation of optical links. In addition, there is a growing interest in the use of LEDs in the visible range, with a dual function that would be obvious and advantageous to provide an efficient lighting in semiconductors and optical communications to achieve [TAN 00].

Table 8.1 shows the different comparative characteristics of the LED and the LD.

Parameters	LED	LD
Spectral width	25–100 nm (10–50 THz)	$<10^{-5}$ to 5 nm (<1 MHz to 2 MHz)
Baseband modulation	Tens of kHz to tens of MHz	Tens of MHz to tens of GHz
E/O efficiency	10–20%	30–70%
Cost	Low	Low to moderate

Table 8.1. *Comparative LED–LD*

For a significant energy flow, it is possible to use one or more lasers, coupled or not with doped fiber amplifiers (power increase), and one or more diffusers (homogeneous distribution of the transmission power). The realization of this configuration seems easier in the spectral window at 1,550 nm because the elements offer this level of power are commercially available.

8.2.3. *Optronics receivers*

Because of their small size, the optronic receiving devices are essentially junctions called photodetector photodiodes (PD), and three main parameters determine the choice in an optical wireless system:

– The response time of a photodiode is characterized by its rise or fall time, the latter being longer. Generally, the time required to reach 90% of the final current of the photodiode. The response time value is important in selecting a photodiode, when this value is low the photodiode detects modulation frequencies more quickly, but its active surface is smaller.

– The sensitivity is the second parameter defined as the minimum received optical power for a correct detection of the transmitted signal. The value of the sensitivity is involved in the selection but it is best to consider the sensitivity of the complete optical reception (see Chapter 6).

– The receiver sensitive and active area is directly proportional to the cosine of the field of view and inversely proportional to the response time.

It is possible to distinguish four main groups: the photovoltaic cell, the PIN photodiode, the avalanche photodiode (APD), and the metal–semiconductor–metal (MSM) photodiode.

8.2.3.1. *Photovoltaic cells*

A photovoltaic cell is the most basic electronic semiconductor component that operates as a battery, produces a current proportional to the incident light when connected to a short circuit. The electron–hole pairs created by the photon absorption are dissociated by the permanent electric field through the space charge region at the P–N junction (positive–negative), interface between the P- and N-doped layers. In addition, in the case of the binary, ternary, and quaternary III–V compounds, the energy gap of the semiconductor can be matched with the photon energy (Figure 8.10).

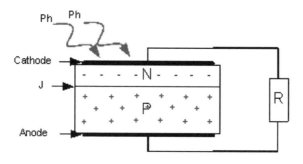

Figure 8.10. *Photovoltaic cell*

Specifically, the top layer of the cell is an N-doped layer whose electron density is very high and thin to let the largest number of photons get through into the junction. The bottom layer of the cell is a P-doped layer whose hole density is very high.

Under the internal electric field action, the free electrons and holes are swept out of the junction area and a space charge area develops around the junction plane. As represented in Figure 8.10, a conductive thin film covers the upper surface (type N: cathode) of the chip to allow the passage of photons. The underside of the opposite chip is deposited on p-type electrode (anode).

The photovoltaic cell can be considered as the combination of an electric generator and a diode. We can add multiple layers and junctions in order to optimize the energy spectrum of photons. We obtain multijunction cells.

This light sensor is not the most suitable modulated signal receiver because its response time is high, in the order of milliseconds. It can be used as an energy resource. The current yields of around 20% are likely to evolve rapidly.

8.2.3.2. *PIN photodiode*

The PIN photodiode is a multilayer structure (basically three layers), the intrinsic I-layer between P-doped and N-doped layers. Under a reverse bias, it operates as a current source, the absorption of photons creates electron–hole pairs which are separated under the high biasing electric field and recombine to provide the photoelectric current. This current has an amplitude proportional to the radiant optical power incident on the active surface.

In general, this settled in the junction of the photodiode reverse current (equation [8.3]), expression (see Chapter 6):

$$i = i_s + i_d$$

where

— i_d is the dark current.

— $i_s = R.P$ photocurrent proportional to the incident radiant power P.

— R is the responsivity.

Equation 8.3. *Current of the photodiode*

The photodiode has a responsivity that depends on the semiconductor material and furthermore on the incident signal optical spectrum (Figure 8.11).

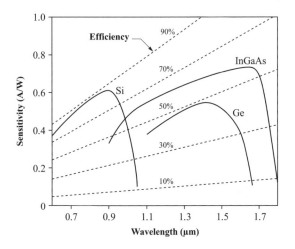

Figure 8.11. *Responsivity versus wavelength*

Material example and sensitivity band:

– Silicon (Si): 400–1,000 nm;

– Germanium (Ge): 800–1,600 nm;

– Indium callium arsenide (InGaAs): 800–1,800 nm.

The PIN photodiodes are widely used due to their proper responsivity, their good tolerance to temperature variations and the low bias voltage. The response time and the active surface are suitable for moderate bandwidth optical wireless communications.

8.2.3.3. Avalanche photodiode

Under high voltage reverse bias, below breakdown, the P–N junction may operate in the avalanche mode with internal multiplication gain. The radiant power of the amplitude-modulated incident signal is often very low; it may be appropriate to amplify the photocurrent. The principle is a chain ionization by impact of the carriers under the influence of an electric field much more intense than that of a PIN photodiode. Each high-energy primary carrier will therefore release several secondary carriers from their ions. The multiplication gain M can be derived from the following equation:

$$i_s = M.R.P$$

Equation 8.4. *Multiple primary photocurrent proportional to the incident radiant power*

In the following expression of the signal/noise power, assuming for simplicity that the quantum noise emission is negligible compared to other sources of noise, we consider the two terms in the denominator:

$$\left(\frac{S}{B}\right)_{PDA} = \frac{(M.R.P_{incident})^2}{2e.(R.P_{incident} + I_{dark}).B.M^2.F(M) + 4kT.B/R_{load}}$$

where

– e is the electron charge.

– k is the Boltzmann constant.

– T is the temperature in Kelvin.

– $P_{incident}$ is the incident optical radiant power.

– R_{load} is the load resistance of the diode.

– The radiofrequency bandwidth and the multiplication factor are, respectively, represented by the letters B and M.

– R is the intrinsic responsivity of the detector at unity multiplication gain.

In the case of transmissions at high data rates, looking into the denominator, we have the possibility to adjust the gain by adjusting the reverse bias voltage applied so that the first noise term becomes equal to the second thermal noise term.

8.2.3.4. Metal–semiconductor–metal (MSM) structure

The MSM structure can also provide improved detection performance by the possibility of obtaining a large active surface with a very low response time and easy implementation. These photodiodes have been the subject of study [TEM 96], due to a simple and compact structure facilitated by a monolithic integration of field effect transistors. Indeed, the sensitivity of the detectors is inversely proportional to the response time which is directly proportional to the capacitive effect of the active surface detection. The MSM detector could be optimized on the basis of a low capacitance per unit area. Experiments have shown beneficial first results [OBR 05] but further studies are needed in this direction.

8.3. Optics

8.3.1. Transmitter optical device

The problem of optical emission in limited space is the following: how can we obtain simultaneously a significant divergence of the transmitted beam (typically ±45° or 90°) and keep a Safety Class 1 with an acceptable radiant power attenuation and a low-volume optical device?

As far as optical emission in free space is concerned, the constraint is less important and can be summarized as follows: by what means can we get a low divergence of the emitted beam, typically a few milliradians, while in a Safety Class 1 with an optical acceptable budget and reduced volume of the device?

The resolution of this problem is based on two main parameters (Figure 8.12): the distance, called focal distance (D) between the emission source (Opt) and the combination of lenses and/or mirrors, and the characteristics of these lenses and mirrors (diameter, radius, etc.).

In free-space optics, several solutions are possible, a small Cassegrain telescope with a combination of diverging lens and diffuser. The first solution is the use of multiple mirrors (Cassegrain telescope). The Cassegrain telescope (Figure 8.13) consists of two mirrors: a concave and parabolic mirror (called primary) and a hyperbolic convex mirror (called secondary). The main advantage of the Cassegrain is its low cost and its limited divergence.

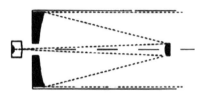

Figure 8.12. *Optical emission*

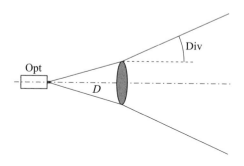

Figure 8.13. *Cassegrain telescope*

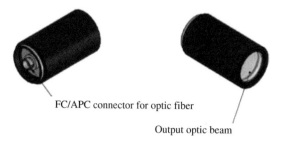

FC/APC connector for optic fiber

Output optic beam

Figure 8.14. *Example of optical transmitter*

8.3.2. *Receiver optical device*

The problem of propagation in limited as well as in free space at the receiver level can also be summarized by one question: what must we do to obtain a high concentration of the transmitted beam in a solid angle as large as possible, typically

±45° or 90°, while avoiding defocusing with an acceptable optical budget through a low-volume optical device, knowing that beyond an angular limit, the incident beam does not reach perpendicularly the active surface of the photodiode?

As far as propagation in free space is concerned, the solution depends on three main parameters: the distance, called focal distance (D) between the combination of lenses and/or mirrors (with their special characteristics) and the photodiode, the characteristics of these lenses or mirrors (diameter, radius, etc.), and the active surface of the photodiode.

Without going back to the calculation mentioned in Chapter 6, it is nevertheless possible to consider several solutions from the Cassegrain telescope to the fish-eye lenses.

In the field of optics in free space, the first solution is a combination of mirrors. As mentioned above, the Cassegrain telescope (Figure 8.13) consists of two mirrors: a concave parabolic (primary) and a convex (secondary) mirror. The main advantage of the Cassegrain telescope is its low cost with a gain (area ratio) and the large field of view is limited to receiving, but the secondary support mirror can cause a very damaging gray area.

In the field of optics in limited space, the first solution is based on a spherical lens. In the research for a signal collection while limiting the length of the device, it is necessary to obtain a ratio of diameter of the lens/focal length as high as possible, but to avoid that, the focus point will move away from the optical axis; it is essential to find a good compromise between the size of the collector surface and a short focal length.

We must therefore move toward a lens as large as possible with the shortest possible focal length. Now, shorter the focus, more the radius of curvature approaches the diameter of the lens, the lens tends to a hemisphere. But to limit spherical aberrations inherent to conventional spherical lenses, we should choose an aspherical lens (Figure 8.15(a)) with a refractive index n. With this approach, the size is smaller with a larger field of view, but due to the thickness of the lens, the absorption signal is more important and its gain is lower (about 3 dB). This approach remains the most usual.

Compared to the problem of thick aspherical lenses, Fresnel lenses are a good alternative. The Fresnel lens (Figure 8.15(b)) is a special type of lens invented by Augustin-Jean Fresnel. Its design allows the achievement of the large field of view with a short focal length, a reduced weight and volume compared to the implementation of conventional lens.

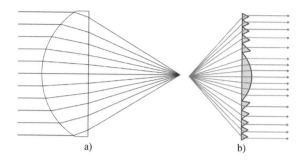

Figure 8.15. *Example of aspherical lens (a) and Fresnel (b)*

In the case of focusing by a Fresnel lens, the area considered is that of the lens also weighted by the recovery of surfaces of the focal spot and the detector surface. The gain is proportional to the ratio of surfaces and a Fresnel lens is not fundamentally different from a conventional lens, except that it is done for a specific wavelength. Fresnel lenses can produce optical devices especially at lower costs (less material is needed) and with a smaller size. Limited divergence and the problem of the orientation of these devices, however, remain the same as that encountered with conventional lenses.

Another possibility is to increase the reception sensitivity using an optic amplifier concentrator or "fireball" [MIH 07]. That is to say, a photomultiplier with microspheres from a part of the energy received, with an adjustable amplification. But this is currently experimental.

Still with the aim to capture the maximum of the incident optical beam with great freedom of incident angle, we should mention the combination of lens for forming the Maxwell "fisheye" [MAX 54]. The "fisheye" device (Figure 8.16) is a lens with an important refractive index to mimic the characteristics of a fish eye and get a wide field of view with defocus limited sensitivity. The size and cost of such a device remains a point to be refined. A combination of "fisheye" and "cat eye" lens provides superior performance in optical terms [ALH 06].

Each solution has advantages and disadvantages and a phase of economic materials integration is essential to reduce the cost and the volume.

It must be remembered that an increase in the field of view also leads to an increase in the optical noise and possible interference. Optical filtering is required to reduce this noise. There are different techniques: the spatial filter (diaphragm) or wavelength filter (attenuator). These are discussed below.

IR Fisheye Lens; Focal length = 3.391; NA = 0.7143; Units: mm

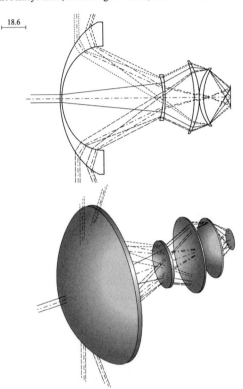

Figure 8.16. *Fisheye lens [ALH 06]*

8.3.3. *Optical filtering*

8.3.3.1. *Spatial filter or diaphragm*

The spatial filter (or diaphragm) is a thin screen with an aperture in its center. The role of this diaphragm is to restrict part of the optical beam to limit the optical noise and avoid saturation of the receiving device. The saturation of the optical detector may be avoided using feedback control of the aperture of this diaphragm. Other electronic solutions are possible such as automatic mitigation (automatic attenuation control – AAC) or automatic gain control (AGC).

8.3.3.2. *Wavelength filters or attenuators*

Ambient light is the largest source of interference or noise and may affect an optical link (see Chapter 4). It induces a constant current on the photodetector of the receiving optical system and can be treated by the electric part of the receiving system [PHA 98].

However, different studies on this subject have been undertaken and the most relevant seems the addition of an optical filter to reject background radiation out of band signal, thereby reducing the incident radiant power level (Figure 8.17). Two main types of filters are presented:

– Absorption by an optical filter can reject particular ambient light and the Sun. It has a wide field of view (theoretically independent of the incident angle) and a filter typically of several hundred nanometers (broadband spectrum). It is possible to appropriately filter by the incorporation of specific layers directly on the photoelectric detector [OBR 00]. The use of a holographic diffuser also allows any filtering of ambient light [JIV 01]. These absorption filters allow us to have band-pass, high-pass, or low-pass filters. Figure 8.17 shows an example of high-pass filter (dashed line) to absorb some solar radiation (black line) while letting the wavelength of the incident beam (bold line at 1,550 nm).

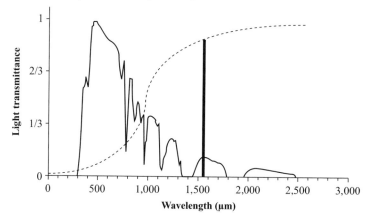

Figure 8.17. *Example of optical high-pass filtering*

– Interference filters are band-pass filters. They consist of multiple layers and are much more effective in filtering accuracy of the order of several nanometers. These filters are sensitive to the angle of incidence of the beam (shift of the wavelength of about 1 nm per degree) and are most often used for beams at normal incidence because there is a loss of signal in terms of power as in a departure from normal incidence. Fabry–Perot filters, for example, are tunable in wavelength according to the angle of incidence. This should also include Bragg mirrors and Fabry–Perot vertical cavities based also on the interference. Experimental systems have circumvented this problem in terms of sensitivity by sticking a soft interference filter on the curved surface of a hemispherical lens [KAH 95]. This maintains the narrow spectral window while increasing the field of view but the cost of implementation is high. The latter filter is close to the ideal filter for a wireless optic reception, e.g. a very narrow band-pass filter insensitive to the angle of incidence and with a large field of view.

8.3.4. *Summary*

The search for the optimal solution is defined by the following elements: high-sensitivity receiver, fast response time, and large active surface. The optimal solution is coupled with the optical device which must provide a high concentration for a maximal equivalent surface and a good reception with a large field of view.

Generally, the optical/electrical converter is associated with one or several low-noise amplifiers (TIA – transimpedance amplifier) which increase the received signal and possibly a receiver automatic control gain (AGC or ACG). This command performs an amplification operation with a feedback loop to control the received signal at a level high enough to allow for proper treatment while avoiding saturation.

Other more exploratory tracks can also be considered, such as:

– Fresnel lens in three dimensions in order to increase the field of view;

– nanophotonics for an optimized compactness of the receiver.

Finally, this section shows that the receiving device depends not only on the receiving optronic device but also on the optical device characteristics and must be a compromise between performance, compactness, and cost. This choice also depends on the type of treatment that is performed on the electrical signal. This point is developed in the next chapter (Chapter 9).

Chapter 9

Data Processing

"They're still using obsolete old electrons"
Jeff Hecht, *City of Light*,
Oxford University Press,
Oxford, 1999.

9.1. Introduction

Data processing is a part of the telecommunication systems that, in emission, processes the signal before the digital/analog conversion and, in reception, after the analog/digital conversion (see Figures 8.1 and 8.2).

The analog/digital converter performs a scanning or digitizing operation and transforms a continuous signal (analog) in discrete values (binary) according to the Nyquist–Shannon sampling theorem. This theorem states that the sampling frequency of a signal must be equal to or greater than twice the maximum frequency of this signal. For example, in the sound domain, a human ear can capture up to 16 kHz and exceptionally up to 22 kHz. The associated analog/digital converter must sample the audio signal at about 44 kHz to ensure a good reconstruction in the opposite operation (digital/analog conversion). For a standard audio CD, the sampling frequency is equal to 44.1 kHz.

The signal processing includes operations such as filtering, compression, analysis, prediction, modulation, and coding. We discuss in this chapter the modulation and coding parts in a specific wireless optical configuration. The other items (e.g. phase signal use) are not directly related to the specificity of the wireless optical propagation channel. They are already the subject of many relevant books [HAE 03, GLA 96, ZIE 98].

9.2. Modulation

Optical waves are electromagnetic waves. It is theoretically possible to transmit digital information by amplitude, frequency, or phase change. In the radio domain, many solutions using the three options simultaneously are available to optimize the spectral congestion (spectral frequency width used to transmit the signal) or the bandwidth (signal modulation frequency). In most cases, the bandwidth is equivalent to the Nyquist band or spectral congestion.

Specifically, in an elementary duration T, one or more bits n can be associated with each symbol. It is associated with the symbol number M by the relation:

$$M = 2^n$$

Equation 9.1. *Number of M states symbol*

As an example, the 16-QAM (16-state quadrature amplitude modulation (QAM) or two 4-state amplitude modulations, $M = 16$) allows us to transmit 4 bits per symbol duration ($n = 4$). So, to transmit a bit rate of 100 Mbps, the modulation rate or symbol rate is only 25 Mbauds (a baud is the number of symbols transmitted per second on a modulated signal). The required bandwidth equivalent to the Nyquist bandwidth is 25 MHz.

Because of the wireless optical emission and reception source characteristics (phase detection difficulty, bandwidth limitation inversely proportional to the receiver sensitivity, intersymbol interference, etc.), it is currently judicious to carry information on the amplitude and the position of the transmitted signal [KAH 97]. In wireless optical communication, one of the most effective solutions is to use the intensity modulation (IM) in which the intensity of the transmitted signal or the optical power is directly proportional to the modulating signal (direct detection – DD). The photodetector current is proportional to the intensity of the received optical signal, which for IM is also the original modulated signal (IM/DD).

9.2.1. On-off keying (OOK) modulation

The on-off keying (OOK) modulation is the simplest form of amplitude shift keying. A binary "one" means an impulse for a specific duration, a binary "zero" means no impulse.

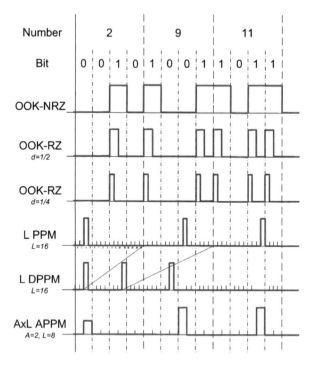

Figure 9.1. *The different modulation techniques*

This OOK modulation (see Figure 9.1, binary sequence) provides two main types of impulsion pulses: non-return-to-zero (NRZ) or return-to-zero (RZ).

For NRZ-OOK modulation, the duration of a pulse is equivalent to 1 bit.

For RZ-OOK modulation, the duration of a pulse is shorter than the duration of 1 bit. This pulse can have different duty cycle values, called "*d*" (duty cycle). When *d* is less than the length unit, the spectral congestion increases with the factor *d*, while the mean power at the same amplitude decreases by an equivalent factor (Table 9.1).

It should be noted that in the amplitude shift keying (ASK) modulation, the amplitude difference distinguishes the binary "zero" from the binary "one" because the binary "zero" is not zero amplitude. It is particularly used in the latest IrDA protocol version in 2-ASK at 512 Mbps and 4-ASK at 1 Gbps. This approach allows an increase in the number of bits for one same symbol period but to the potential detriment of the optical link margin.

OOK modulation is a particular case of ASK modulation where the binary "zero" amplitude is equal to zero.

To limit the "zero" or "one" bit sequence which may cause a synchronization loss or a disconnection (a bits sequence of "zero"), a simple coding is usually added to prevent the DC component. This coding is, for example, the "Manchester" or the 4B/5B coding. The maximum efficiency is achieved by the RZ-OOK modulation with $d = 1/2$. When the duty cycle d is equivalent to 1/2, the efficiency becomes identical to a 2-PPM modulation.

The dotted lines between L-PPM and L-DPPM represent the first bit of the next number.

9.2.2. The pulse position modulation

The pulse position modulation (PPM) is based on the pulse position to transmit information and is divided into two principal categories (binary sequence shown in Figure 9.1 with $L = 16$):

PPM with L options (L-PPM): It is to transmit binary information by the position of the pulse on L options. The mean power of the signal in 1/M is lower, but the bandwidth increases by an equivalent factor.

Differential pulse position modulation (DPPM): The position modulation is differential. This modulation improves the ratio between the power and the bandwidth. In addition, the synchronization detection is not necessary because we have a differential position modulation [SHI 99].

Other variants are also available such as the digital pulse interval modulation (DPIM) or the overlap pulse position modulation with L options (L-OPPM). There are also modulations which offer a combination of the amplitude and the position, such as amplitude pulse position modulation (AxL-APPM) families. An example of a binary sequence with two amplitude levels and eight position options is presented in Figure 9.1.

Generally more complex to implement, this modulation family is considered as the best technique for the line of sight optical communication systems (IM/DD systems) due to a better efficiency of the mean power.

The PPM is used in many optical systems and has been adopted by IEEE 802.11 [IEE 11b] and IrDA groups as the standard for the first proposed protocols in the physical layer [IRD 11].

But for wireless optical communication in a confined environment, this solution becomes less effective for throughputs greater than 100 Mbps with direct consequences such as the problems of synchronization and the increase of the sensibility to intersymbol interference [KAH 95, AUD 96].

A comparison of the modulation schemes is presented in Table 9.1 in function of mean power and bandwidth for some examples related to Figure 9.1.

Modulation	Mean power	Mean power (example)	Bandwidth	Bandwidth (Throughput = 10 Mbps)
OOK-NRZ	$\dfrac{P}{2}$	$\dfrac{P}{2}$	BP	10 MHz
OOK-RZ $(d = 1/2)$	$\dfrac{P}{2d}$	$\dfrac{P}{4}$	$2BP$	20 MHz
OOK-RZ $(d = 1/4)$	$\dfrac{P}{2d}$	$\dfrac{P}{8}$	$4BP$	40 MHz
L-PPM $(L = 16)$	$\dfrac{P}{L}$	$\dfrac{P}{16}$	$16BP$	160 MHz
L-DPPM $(L = 16)$	$< \dfrac{P}{L}$	$< \dfrac{P}{16}$	$16BP$	160 MHz
AxL-PPM $(A = 2, L = 8)$	$\dfrac{P(1 + A)}{2L}$	$\dfrac{3}{16}P$	$8BP$	80 MHz

Table 9.1. *Mean power and bandwidth*

9.2.3. *The orthogonal frequency-division multiplexing (OFDM)*

The orthogonal frequency-division multiplexing (OFDM) technique is a method for digital signals coding using multiple orthogonal frequencies [WEI 71, CIM 85]. The principle is a distribution of digital data on a large number of subcarriers. To limit the spectrum congestion, the subcarriers overlap but, due to their orthogonality, they do not interfere with each other.

Each subcarrier is independently modulated using conventional digital modulations. They contain several binary information per symbol time, such as

QAM or frequency shift keying (FSK). This feature added at the insertion of a guard interval after each transmitted symbol provides good resistance to intersymbol interference at the reception.

The separation of these carriers is carried out using a fast Fourier transform (FFT).

The discrete multitone (DMT), the coded orthogonal frequency-division multiplexing (COFDM), and the orthogonal frequency-division multiplexing/ offset quadrature amplitude modulation (OFDM/OQAM) are based on the same principle. The last two introduce a code per channel and increase the process efficiency [LEF 95].

The success of this process has led to its use in many areas. This extends from wireless networks (Wi-Fi) to broadband digital radio communication (DRM), or from digital broadcasting (DVB) to copper connections (ADSL).

For wireless optical communications in confined environment, the study of DMT, OFDM/OQAM, and COFDM can offer very high data rates in comparison to OOK modulation (a ratio from 3 to 10) over longer distances (over 30%) [OME 10a].

The modified COFDM/OQAM even presents an important advantage in comparison to COFDM in terms of data rate and distance [ELT 09]. But for the latter case, the obtained benefits should be put into perspective because it is a simulation, but significant gains are still possible.

9.2.4. *The diversity: MIMO*

Multiple-input multiple-output or MIMO is a technique of signal processing for wireless equipment. It offers improved performance of communication systems using multiple inputs at the emission (antennas) and multiple outputs at the reception (antenna) for the same elementary link (view from the propagation channel). These rate and coverage performances are improved in the same bandwidth and with the same power levels. This is achieved by increasing the spectral efficiency (number of bits per second per hertz (bit/s/Hz)) due to the diversity and spatial multiplexing.

The classical device is the single-input single-output (SISO) communication system and then several variations are possible (Figure 9.2).

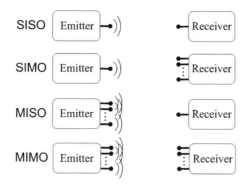

Figure 9.2. *The different input–output techniques*

The first ideas were proposed by Kaye and George in the early 1970s and developed as a patent by Winters and Salz at Bell Labs in 1984. MIMO technology has attracted much attention in the field of wireless communications, since the MIMO channel capacity increases proportionally to the number of antennas.

MIMO technology can be defined in three parts:

Preconfiguration: After having read the channel state information (CSI), the emitter defines the emission characteristics of each antenna (beamforming) to optimize resources and to reduce selective fading.

Spatial multiplexing: Spatial multiplexing allows the use of the same frequency to send different data on each of the antennas and this increases the global throughput. The number of streams is limited by the number of transmitting or receiving antennas and an estimation of the channel state information is not necessary,

Coding diversity: This approach is also used when there is no channel state information estimation and therefore no optimization of the stream on antennas. On each channel, the signal is coded using techniques called space-time coding. This signal, emitted from each transmission antennas, uses coding diversity and selective fading independently of each of the antennas to increase the diversity of the signal.

The MIMO technique is often associated with OFDM coding. It is part of the IEEE 802.16 and the IEEE 802.11n standards. Several solutions are already commercially available, especially in "box" and Wi-Fi systems.

An evaluation of MIMO technology to be used in optical wireless communications is the subject of several studies and a multispot prototype [OME 10b] with significant throughput (1 Gbps) and coverage (three times) gains. In multispot diffuse (MSD), several studies have been performed [ALQ 03b, ALQ 04, JID 04] without notable experiment.

9.2.5. *Summary*

In wireless optical communications, the ideal solution must offer a high throughput in a limited bandwidth to have a suitable optical link budget. In a confined environment and from a few hundred Mbps, it must also have good intersymbol interference resistance.

CODFM, OFDM/OQAM, and ASK approaches are relevant with a number of subcarrier or limited levels to not prejudice the optical link budget while obtaining a reduced bandwidth.

NRZ-OOK modulation can provide, with difficulty, very high data rates with a limited bandwidth. But when it is combined with a suitable equalization at the receiver [MAR 96], it is possible to increase the data rate with limited bandwidth, and this is in a way comparable to COFDM or COFDM/OQAM solutions. But the use of these equalizers can greatly increase the complexity of the receiver [GON 05].

A MIMO approach has yet, been little tested, in the wireless optical domain in a confined environment but it seems that the performance (throughput and coverage) are usually similar to that of a radio system.

In addition, experimental advances [BIG 10, BEL 10] and the availability in the medium term of a coherent detector in the optic domain will offer an increase in sensitivity performance, in compactness, and will eventually allow the use of the signal phase component to transmit information.

The selection of the optimal solution must also take into account the parameters such as complexity/cost and benefits obtained in an integration constraint extended to optical and optronic modules.

9.3. The coding

9.3.1. *Principle and definitions*

9.3.1.1. *Principle*

In 1948, Shannon laid the foundations of information theory in his famous article [SHA 48]. It provides laws on the optimal compression of digital information or source coding, which aim to eliminate redundancy in this information to better exploit the available rate. A perfect compression makes equiprobable the emission bits representative of the information.

Shannon also describes the conditions for achieving reliable transmission, regardless of the considered medium, by introducing the concept of channel capacity. He introduces the main theorems of error correction coding or channel coding and shows that it is always possible to perform a transmission with an arbitrarily small error probability, since the transmission rate is less than the channel capacity. This is made possible through the use of an appropriate error correction code, which adds redundancy information to be transmitted. This redundancy is exploited at the reception to detect or even to correct errors.

The error correction code is not explicitly given by Shannon, but its existence is theoretically guaranteed. Therefore, much research has been conducted to find powerful error correction codes. Among these codes, there are several main categories: basic code, block codes, and convolutional codes whose construction, properties, coding, and associated decoding algorithms are widely described in several publications [MIC 85, GLA 96].

The use of iterative processes in the channel coding has enabled the decoding of performing error correction code while maintaining a reasonable computational complexity. The introduction of convolutional turbo codes by Berrou and Glavieux in 1993 [BER 93], the block turbo codes by Pyndiah *et al.* in 1994 [PYN 94], or low-density parity-check coding (LDPC) by Gallager in 1963 [GAL 63] allowed the achievement of decoding performance near the Shannon limit in the Gaussian case.

9.3.1.2. *Definitions*

The error correction coding consists of adding redundancy to the transmitted signal (as opposed to compression). The use of this redundancy at the reception improves the transmission quality to the detriment of useful data throughput, the same information in different forms being issued to ensure its protection. A correction code allows the detection of errors in the simplest cases and error correction for more complex systems.

9.3.1.2.1. Coding efficiency

A linear coding operation f_c consists of associating with the message m composed of k information elements, a code word $c = f_c(m)$ composed of n elements containing the transmitted information and redundancy.

We define the coding rate (equation [9.2]) as the ratio between the number of data elements and the number of elements contained in a code word, i.e.:

$$R_c = \frac{k}{n}$$

Equation 9.2. *Coding efficiency*

For example, a device with an Rc encoding rate equal to 0.6 indicates that 40% of the transmission is assigned to the coding.

9.3.1.2.2. Correction power and code detection

The minimum distance of a code allows us to define the detection power, denoted by u, by the relation $d = u + 1$. So any received sequence with less than u erroneous elements can be detected in reception.

The minimum distance of a code also sets the power of correction, denoted by t, by the equation $d = 2t + 1$. So any word containing less than t erroneous elements can be fully corrected in reception because there will be no ambiguity on the closest code word.

The combination of these two powers has the potential to detect and correct c errors on m detected errors, using the relationship $d = m + c + 1$.

9.3.2. *Example of coding*

9.3.2.1. *Basic codes*

9.3.2.1.1. The parity check

The parity check (vertical redundancy check – VRC) is one of the simplest control systems and the easiest to implement. This results in the addition of a bit (parity bit) to a number of data bits (code word). The best known example is the addition of a bit to a 7-bit word to form a byte. The value of the parity bit (0 or 1) detects an error in the sent byte. For example, the parity bit is "one" if the number of bits equal to 1 in a code word is odd, or "zero" otherwise.

By extension, the cross-parity check or longitudinal redundancy check (LRC) performs a check on the parity bits of a block of characters.

9.3.2.1.2. The cyclic redundancy check

The cyclic redundancy check (CRC) is a powerful means to control data integrity and also easy to implement. It is the main error detection method used in telecommunications.

The CRC directly protects data blocks or frames. A block of data or control code (CRC) is integrated in each frame. This code contains redundant elements in the frame and can not only detect errors but also repair them. The principle of the CRC is to treat binary sequences as binary polynomials, e.g. polynomials whose coefficients correspond to the binary sequence.

This predefined polynomial is known by the transmitter and the receiver. At the emitter, a CRC is generated on the bits of the frame using the algorithm. These two elements are transmitted to the receiver that performs the same calculation to verify if the CRC is valid.

The most commonly used generator polynomials are:

– CRC-12: $X^{12} + X^{11} + X^3 + X^2 + X + 1$

– CRC-16: $X^{16} + X^{15} + X^2 + 1$ (This code is used in the 802.11 IR PHY standard.)

– CRC CCITT V41: $X^{16} + X^{12} + X^5 + 1$ (This code is particularly used in HDLC procedure.)

– CRC-32: $X^{32} + X^{26} + X^{23} + X^{22} + X^{16} + X^{12} + X^{11} + X^{10} + X^8 + X^7 + X^5 + X^4 + X^2 + X + 1$ (This code is implemented in the Ethernet standard.)

– CRC ARPA: $X^{24} + X^{23} + X^{17} + X^{16} + X^{15} + X^{13} + X^{11} + X^{10} + X^9 + X^8 + X^5 + X^3 + 1$

9.3.2.2. *Block codes*

Among the block codes, we consider the Bose–Chaudhury–Hocquenghem binary cyclic codes (BCH codes) [BOS 60] and Reed–Solomon q-ary codes (RS codes) [REE 60]. These codes are called algebraic because they are built on a polynomial extension of a Galois field. The encoding operation of algebraic codes can be done using registers, making it a popular code in terms of practical implementation and speed of execution, and this, even at high rate.

A classical decoding method of cyclic codes uses the syndromes method [MIC 85]. However, an efficient decoding of BCH or RS codes operates their algebraic properties. Various algorithms with robust input–output (values 0 or 1) exist such as Peterson's algorithm [PET 72] and Berlekamp–Massey's algorithm [BER 68].

We retain the Berlekamp–Massey algorithm whose complexity varies linearly with the correction power t of the considered code. On the implementation point of view, Peterson algorithm is a highly effective alternative for correction power t equal to 1 (or $t = 2$) and gives exactly the same performance as the Berlekamp–Massey algorithm.

9.3.2.2.1. BCH codes

BCH codes are built on a polynomial extension of order m of a Galois field denoted by GF (2^m) [PAP 95]. They are denoted by BCH (n, k, d), where n is the number of bits of a code word, k the number of bits of the encoded message, and d the code binary distance. We have the following relations:

$$BCH(n,k,d) \Rightarrow \begin{cases} n = 2^m - 1 \\ n - k \leq mt \\ d \leq n - k \end{cases}$$

Equation 9.3. *BCH codes*

If we consider a series of BCH codes which possess a correction power $t = 1$, these codes being binary, so they can correct up to a false bit among all the bits of an m-length code word.

Figure 9.3 shows examples of curves of probability of bit error rate (BER) on a Gaussian channel of an NRZ-OOK modulation [LAL 07] associated with various BCH codes whose power correction is $t = 1$ bit in function of electrical signal-to-noise ratio (SNR; RSBe): solid lines (simulated curves) and dotted lines (theoretical curves). The theoretical bounds are also given and we can see that the simulated and theoretical curves overlap.

Note that the gain provided by BCH coding is modest compared to a non-coded transmission. For a BER of 10^{-9}, these (electrical) gains range from 2 dB for the code BCH (255, 247, 3) with performance Rc = 0.9686 to 2.5 dB for the code BCH (15, 11, 3) with performance Rc = 0.7333. The more reduced is the coding efficiency (and thus the effective transfer rate), better is the performance.

The selected codes are easy to decode, but not very powerful ($t = 1$). BCH codes with greater distance would gain more in spite of the coding efficiency and the decoding facility. For a given distance, greater is the BCH codes, size (and thus the size of the associated Galois field), greater is the coding efficiency. Choose more powerful BCH codes, while maintaining a high coding efficiency use longer BCH.

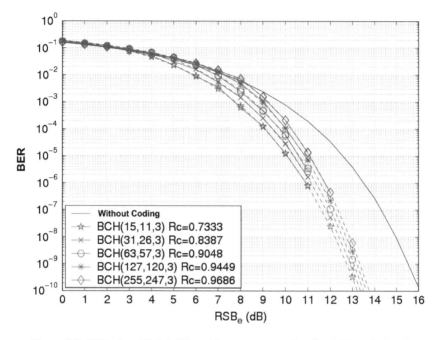

Figure 9.3. *BER of an NRZ-OOK modulation associated with a BCH code (t = 1)*

9.3.2.2.2. RS codes

RS codes are also built on a polynomial extension of order m of a Galois field denoted by GF (2^m). However, unlike the binary BCH codes, these codes are q-ary codes, where $q = 2m$. They are denoted by RS (n, k, d), where n is the number of q-ary symbols of a code word, k the number of q-ary symbols of the coded message, and d the code q-ary distance. We have the following relations (equation [9.4]):

$$RS(n,k,d) \Rightarrow \begin{cases} n = 2^m - 1 \\ n - k = 2t \\ d = n - k + 1 \end{cases}$$

Equation 9.4. *RS codes*

Consider a set of RS codes with a correction power equal to $t = 1$. Thus, the selected codes can correct an erroneous q-ary symbol among the n q-ary symbols of a code word. As we can associate with any q-ary symbol only m bits packet, the RS codes can correct errors on bits packets. For example, RS (255, 239, 17) code can correct up to eight consecutive bits if they belong to the same q-ary symbol ($m = 8$, $n = 255$, $t = 8$, $k = 239$).

Figure 9.4 shows different curves of the probability of BER of an NRZ-OOK modulation associated with RS codes with a correction power $t = 1$ q-ary symbol on a Gaussian channel (LOS) in function of electrical SNR: solid lines (simulated curves) and dotted lines (theoretical curves). The theoretical performance of the code RS (255, 239, 17) ($t = 8$), widely used in current systems, is also plotted.

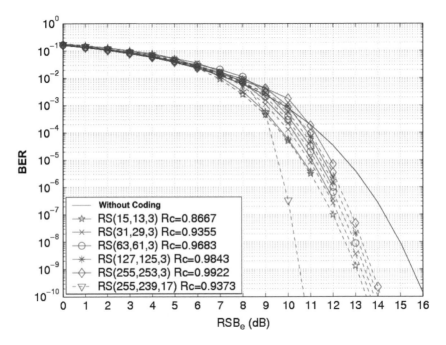

Figure 9.4. *BER of an NRZ-OOK modulation associated with an RS code*

We see the same level of gain (in the range of 2–2.5 dB for a BER of 10^{-9}) with RS and BCH codes compared to a non-coded transmission. Nevertheless, RS codes have the property of maximum separable distance ($d = n - k + 1$) and therefore offer the highest efficiency for a given distance. This is not the case for BCH codes which propose lower efficiency for the same distance as their RS counterparts.

We note finally that the traditional RS (255, 239, 17) code provides an (electrical) gain of 5 dB at a BER of 10^{-9}, for an efficiency of Rc = 0.9373, which makes it particularly attractive.

The slope of the theoretical curve is much more abrupt than that of other RS-presented codes. It is naturally explained by a higher correction power. This code corrects up to 1 byte (or 8 bits packets) among the 255 bytes of a code word. Its use is widespread due to its excellent properties, the correspondence between a q-ary symbol and a byte, which is very practical to implement and the speed of decoding, about 2 µs at 200 MHz.

9.3.2.3. Convolutional codes

Another large family of error correcting codes is represented by the convolutional codes. To calculate each block of n bits, a convolutional encoder uses not only the block of k bits that it has in input but also the previous m blocks of k bits that have already encoded, thus introducing a memory effect. We call the quantity $\mu = m + 1$ the constraint length of a convolutional code. More important is the constraint length, greater good performance the code offers at the detriment of an increased decoding complexity.

The encoding operation uses linear combinations between the received k bits block and the encoder memory. The use of linear operations linearizes a binary convolutional code. Thus, a k zeros block is encoded by a block of n zeros.

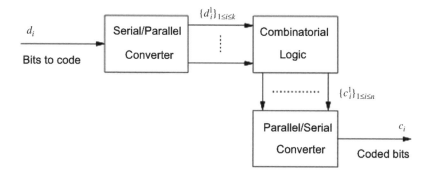

Figure 9.5. *Schematization of a convolutional encoder with performance Rc = k/n*

Combinatorial logic (Figure 9.5) generating the ith bit of the code word $(1 \leq i \leq n)$ from k $(m + 1)$ bits to be encoded is represented by a polynomial g_i. The term "convolutional" comes from the calculation of the encoder output, which is then equivalent to a convolution product between the bits to encode and the polynomials g_i of the encoder.

We consider convolutional codes with $k = 1$ and it is denoted by CC $(g_1, ..., g_n)$. The convolutional code of performance Rc $= 1/n$ is defined by the polynomials g_i. Considering the convolutional code CC (7.5) where $m = 2$, $g_1 = [1\ 1\ 1]$, and $g_2 = [1\ 0\ 1]$, the coding scheme is shown in Figure 9.6.

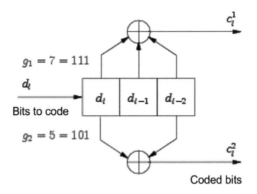

Figure 9.6. *Example of a schematic diagram of the convolutional encoder CC (7.5)*

The use of the m previous input in the encoding operation is used to define a 2^m states trellis representation from the convolutional code. The sequence of bits in a coded message can be represented by a path in this trellis and the decoding is defined as the search for the most probable path. Among the convolutional codes, decoding algorithms are the standard Viterbi algorithm [MIC 85] and BCJR [BAH 74]. The Viterbi algorithm provides definitive output (values 0 or 1) from definitive or soft input (values between 0 and 1), while the BCJR provides soft outputs from soft inputs.

The Viterbi algorithm is widely used in communications, such as Global System for Mobile (GSM) communications.

In the case of the CC (7.5) code, the trellis has only four states, making it a simple code to decode. This code has a free distance $d_f = 5$. The free distance is the smallest Hamming distance between two paths that diverge and converge again. Like the Hamming distance of a code block, the free distance characterizes the performance of the convolutional code. Longer is this free distance, more powerful is the considered convolutional code at the detriment, there again, to an increased decoding complexity.

Figure 9.7 shows the probability of error for a Gaussian channel of NRZ-OOK modulation associated with the convolutional code CC (7.5) of Gaussian channel (LOS) in function of the electrical SNR. A convolutional code being very sensitive to occurring packet errors, we insert a conventional interleaver at the output of the encoder and a deinterleaver before the decoder. It was considered, here, a pseudorandom interleaver of 2,048 bits. The decoding algorithm used is the Viterbi algorithm with definitive inputs and outputs (values 0 or 1).

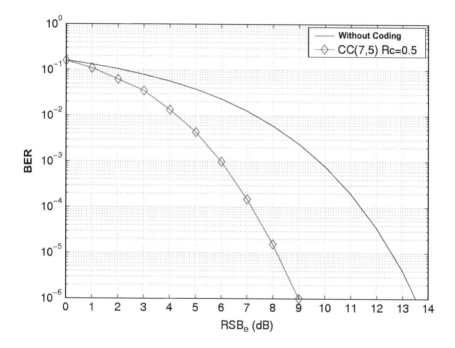

Figure 9.7. *Schematic example of error probability of an NRZ-OOK modulation*

With an error probability of about 10^{-6}, it is possible to obtain a gain of 4.5 dB in the coded version from the non-coded version. If this gain is greater than that provided by the block codes, it must be remembered that the performance of convolutional code is very low (Rc = 0.5) compared with performance of presented BCH and RS codes.

Finally, in the field of digital communications, the turbo codes are a recent category of correcting code (year 1990); and the LDPC code has reappeared in the late 1990s. These two codes are currently closest to the Shannon limit and offer high-performance results. In the field of broadband wireless optical communication (minimum 100 Mbps), these codes are currently less suitable because of the relatively high complexity of the decoding and consecutively the significant processing times.

Other techniques were also studied, for example a PPM Trellis encoding has been characterized for multipath and diffuse infrared transmission [LEE 95]. The trellis coding uses a specific partition to separate the positions of neighboring symbols as a method of efficient coding. Electrical gains of 5.0 dB have been reported for a rate 2/3 coded 8-PPM relative to a non-coded 16-PPM with the same bandwidth [LEE 97].

Another example is the run length limited encoding/coding (RLL/RLC)-HHH (1,13) whose acronym is based on the initials of the researchers who invented it. The encoding RLL is a coding system to correct potential errors made by the temporally distorted pulse (ISI) and an imperfect synchronization. Other modulations, such as PPM, DPPM, and OOK, do not directly address these limitations [FUN 02] but an increase in throughput leads also to more limited performance.

9.3.3. Summary

For a wireless optical communication, the ideal solution must provide an efficient detection and correction power without providing a significant complexity and a long processing time, with of course, a performance coding efficiency. The strategy commonly used in communication systems is to carry out, at the emission, a high-performance block coding and a convolutional coding. So in reception, the convolutional decoding will perform the main task of error correction, and the high-performance block code will eliminate the residual errors.

In the field of wireless optical communication at data rates exceeding 100 Mbps, the simplest technique is the Reed–Solomon error correcting code and the most common approach, which approaches the Shannon limit, is the combination of a Reed–Salomon error correcting code and a short length Viterbi convolutional code, also known as RSV. The combination of a block code (RS) and a convolutional code (Viterbi) currently offers a suitable compromise with an (electrical) gain of about 7 dB.

At constant rate, the increase in processing speed of microprocessors (Moore's law – the speed doubles every 18 months) provides insight into the use of turbo codes and LDPC code in the medium term with greater gains.

Finally, it may be wise to recall the existence of a squaring in the optical/electrical conversion (photodetector). This results in a gain G equal to x dB in the electrical part due to the coding process. But this corresponds to a gain of G equal to $x/2$ dB in the optical part. In the example of the preceding paragraph, the electrical gain of 7 dB power is equivalent to 3.5 dB gain in optical.

Chapter 10

Data Transmission

"The most constant characteristic
of computing is the ability of users
to saturate any system available to them"
Paquel Act,
Corollary computer
Murphy's Law, 1947

10.1. Introduction

10.1.1. *Definition*

We now turn to a presentation of the data link layer, the second layer of the open systems interconnection (OSI) and always the first layer of the Department of Defense's system (see Figures 3.10 and 3.11 of Chapter 3). The protocols of this layer process request services from the network layer (layer 3) and carry out a solicitation of requests for services to the physical layer (layer 1). The purpose of these protocols (protocol data unit – PDU) is the transfer of data between two adjacent nodes of a local or wide area network. The functions of these protocols are mainly:

– frame transfer between two local devices;

– addressing equipment delivery with a frame header containing the source address (from the frame) and the destination address (receipt and processing);

– arbitration of access to means of communication;

– possibly a mechanism for receipt (acknowledgment) and validation of received frames or request for retransmission;

– possibly a device for detecting errors;

– possibly an error correction device.

There are many examples of protocol, such as Asynchronous Transmission Mode (ATM), Infrared Data Association (IrDA), or WiFi, but the best known example of a data link protocol is Ethernet (IEEE 802.3). In this example, there are two main sublayers:

– Media access control (MAC) sublayer manages the access to physical media, the creation of frames, the implementation of the addressing of frames, encapsulation, and possibly virtual LAN (VLANs);

– Logical link control (LLC, IEEE 802.2) sublayer manages the interface with higher layer protocols, provides addressing and control, performs data link with possibly detection, and error correction. Several LLC protocols can coexist on the same MAC sublayer.

These protocols are proposed on a computerized form (driver) with the circuit board or module network. The operating system of a computer or a router has a software interface between the data link layer and the network layers located above. The best known example of network and transport layer protocols is Transmission Control Protocol/Internet Protocol (TCP/IP). These protocols deal with such internetwork routing or global end-to-end addressing.

10.1.2. *The access methods*

In the field of communications, performances are related to the technical choices for efficient access to a transmission channel to multiple users. The protocols of the MAC sublayer propose several approaches based on several forms distributed or centralized access with a static or dynamic allocation and time, or frequency or code multiplexing: time division multiple access (TDMA), *frequency division multiple access* (FDMA), code division multiple access (CDMA), carrier sense multiple access (CSMA), wavelength division multiple access (WDMA), and space division multiple access (SDMA).

10.1.2.1. *Time division multiple access*

With this approach, time is divided into intervals. In each of these intervals, each user transmits its signal. The TDMA appears to be effective under the regular flow of information (flow relatively constant); however, it may be more difficult to manage for the so-called "burst" transmission - very high volume over a very short time [PAH 95]. For example, the Global System for Mobile Communications protocol has a static resource allocation in the form of a time interval (time slot) available in the uplink and downlink with a centralized management by the base station.

10.1.2.2. *Frequency division multiple access*

With this technique, the spectral band is divided into portions of frequency bands (channels) allocated to each user. The receiver filters the desired signal. The FDMA solution is effective in applications with relatively constant flow rates [PAH 95]. However, power efficiency achieved by FDMA technique becomes smaller as the number of users increases.

10.1.2.3. *Code division multiple access*

This technique provides a system for encoding data based on spread spectrum. Different users use different sequences of orthogonal codes for their communications so that no interference has occurred [MOR 96]. This approach is one of the standards for 3G mobile services (CDMA2000).

10.1.2.4. *Carrier sense multiple access*

Another approach to managing shared physical access to the network is proposed by the CSMA protocol. Each device listens to the physical connection to see if another device transmits a frame. If this is not the case, suppose that the first equipment can emit. This approach is probabilistic and mechanisms are used to detect potential collisions (multiple frames are sent simultaneously by multiple devices) and avoid these collisions (collision detection (CD) and collision avoidance (CA)).

For example, the first Ethernet protocol (IEEE 802.3) uses the CSMA/CD protocol. Another example is the IEEE 802.11n (WiFi) which provides management with a dynamic and random allocation CSMA/CA.

In the field of optical wireless, it is possible to use optical multiplexing techniques based also on the channel.

10.1.2.5. *Wavelength division multiple access*

This solution is equivalent to the FDMA and widely used in the field of optical fiberes. Using this technique, each transmitter transmits at different wavelengths using narrow-band transmitters, such as laser diodes. The receiver has a band-pass optical filter that extracts the desired wavelength before the detection and treatment processes.

The wavelength tunable laser diodes are currently complex and expensive, but the prospects of treatment for multiuser management, or jump wavelength, remain attractive.

10.1.2.6. *Space division multiple access*

This technique uses an angular diversity device (several transmission or reception elements pointed in different directions) to emit or receive signals from different directions simultaneously.

This approach can optimize the radiation pattern (smart antenna) and thus obtain a more important optical budget. Although SDMA is a multiple-input multiple-output (MIMO) technique, MIMO is not necessarily an SDMA technique.

In addition, duplex term is used to characterize a communication channel and there are three modes of communication:

Simplex: The channel is unidirectional. It transmits information in only one direction, for example infrared remote control.

Half-duplex: Information is transmitted in both directions but not simultaneously and there are two modes:

– Time division (TDD – time division duplex) of the communication channel using the same transmission resource (the radio channel, for example). This approach is advantageous when the transmission and reception flow rates are variable and asymmetric. Thus, if the emission rate increases or decreases, more or less bandwidth can be allocated.

– Frequency division on the same antenna (FDD – frequency division duplex) or wavelength division (WDD – wavelength division duplex). This solution has a capacity more advantageous because it is possible to assign a wavelength in the downlink and one wavelength in the uplink direction. Thus, the system has also better resistance to the optical coupling problem.

Full-duplex: In this mode, information is transmitted in both directions simultaneously, a point-to-point (PtP) Ethernet cable, for example.

10.1.3. *Quality of service parameters*

The quality of a communication can be defined by several parameters. These elements are used to define the quality of service (QoS) of a transmission. Some of these parameters are listed below:

– data rate (physical layer) and data throughput (network layer, IP, for example);

– latency and processing time;

– packet error rate (PER);

– energy consumption.

The following section shows the various protocols implemented inside wireless optical communication systems.

10.2. Point-to-point link

10.2.1. *The remote control*

Since the early 1980s, the vast majority of remote controls use infrared technology (950 nm) and the transmission of orders is performed by transmitting a digital signal whose frequency is modulated at a few tens of kilohertz (some characters for each command). This communication is simplex or unidirectional. There is no access method in the protocols used by remote controls. The optical noise is minimized by encapsulating the optical receiving part in a plastic which is opaque to visible light but transparent to infrared.

On transmission protocols, there are several solutions for encoding information from different manufacturers: Philips RC5, SIRSC for Sony, Denon RC-284, Sanyo VHR-291st, Decca, Toshiba, Scientific Atlanta, NTL, Nippon Electric Company Limited (NEC), Samsung, etc. The simplest systems send a preamble consisting of short (digit 0) and long (digit 1) intervals. The demodulator performs the synchronization of the signal fairly quickly, typically in less than two cycles. This header allows the automatic gain control (AGC) of the receiver to adjust the level of the transmitted signal to the processor, and to manage the interval changes of the following pulses of the control signal. The first type of pulse position modulation (PPM) was invented by NEC. The protocol sends a binary code for one type of equipment (for instance TV or DVD player) and another binary code for the command corresponding to the pressed button. Then a transmission of these two codes in reversed binary. This solution has two advantages: first, the total message length remains constant, and second it provides a double check with a validated code and, immediately after, its inverse.

Another well-known and widely used protocol is RC5 from Philips. This protocol is a set of 2,048 different instructions, divided into 32 addresses, each with 64 instructions. Each device uses its own address, for example to change the volume on the television without changing the volume of high-fidelity equipment. The transmitted frame consists of 14 bits whose distribution is:

– 2 "start" bits for the AGC of the infrared receiver;

– 1 "toggle" bit state change each time a new button is pressed on the IR transmitter;

– 5 address bits of equipment;

– 6 bits of instruction for the requested function (button pressed).

CD/DVD	Legend	Command
18	DISP	Display
00	0	0
...	...	...
09	9	9
59	A-B	
29	RPT	Repeat
58	CLR	Clear
41	PROG	Program
53	|>	Play
54	[]	Stop
32	|>|	Go to start of next track
33	|<|	Go to start of this track
50	|>|>|	Fast forward
52	|<|<|	Fast backward
48	"	Pause
45	OPEN	Open
87	OK	Okay
80	Right arrow	From main circular button
81	Left arrow	
85	Down arrow	
86	Up arrow	
64	PMODE	Play mode
65	SUBT	Subtitle
66	TITLE	Title
67	MENU	Menu
68	ZOOM	Zoom
69	AUDIO	Audio
70	MEM	Memory
72	RTN	Return
74	SETUP	Setup
75	STATUS	Status
76	SRCH	Search

Table 10.1. *RC5 code for DVD reader*

This RC5 code uses two states (RZ) modulation technique, meaning that each bit is composed of two parts that are never identical. Thus, a transition is always an up/down or down/up bit. The RC5 code uses a down/up transition for the binary 1 and an up/down transition for the binary "0". As an example, Table 10.1 presents RC5 code functions for a CD/DVD reader.

10.2.2. *Infrared Data Association*

Established in 1993, the IrDA is an association of about 150 members from different companies. This association proposed, in 1997, a recommendation for a digital communication with cheap optical modules. This IrDA device appears on many portable devices including computers or mobile phones as well as computer peripherals, such as printers and camcorders.

In general, the facilities are equipped with a short-distance infrared port mainly to transfer files. They generally consist of a pair of transmitter/receiver operating at about 850 nm, and offer data rates from 1.5 Kbps to 1 Gbps over a few meters with a divergence of $\pm 15°$.

The access technique is based on a half-duplex PtP TDMA approach. The architecture of the standardized IrDA protocol compared with the OSI layers architecture is presented in Table 10.2.

OSI layer	IrDA layer
7 – Application	*Point and shoot profile* (PnS)
6 – Presentation	*Object exchange protocol*
5 – Session	(OBEX)/IrCOMM/IrLAN
4 – Transport	*Tiny TP*
3 – Network	*Information access service* (IAS)
2 – Link	*Link management protocol* (IrLMP)
	Link access protocol (IrLAP)
1 – Physical	*Infrared controller, transceiver*

Table 10.2. *Comparison protocol OSI versus IrDA*

These layers can be divided into two categories:

– The required layers are the following:

  - physical layer (specifies optical characteristics, the data encoding, encapsulation for different flow rates);

- IrLAP (manages connections between devices);

- IrLMP (handles the multiplexing of information across multiple channels);

- IAS (for exchanging information between applications running on different IrDA devices);

– The optional layers:

- TinyTP (tiny transport protocol, this layer adds a flow control for connection);

- IrOBEX (facilitates the transfer of files and other data objects);

- IrCOMM (serial and parallel port emulation allows existing applications that use the serial and parallel ports to operate without change);

- IrLAN (local area network access allows access to IR laptops and other devices).

Communication is a half-duplex PtP with time division with a primary equipment or coordinator (A) and a secondary equipment (B) which performs a communication session according to the following approach:

– After a request by the user or automatically, the device containing the infrared module A initiates the procedure for detecting a second module containing the infrared device B.

– After detection, communication engages with an agreement on the most appropriate flow rate according to:

- the specific characteristics of the infrared modules of each devices; they can vary even in the range of products from the same manufacturer:

1) the distance, the divergence, and the field of view (FOV);

2) the embedded version of the protocols on each of the two devices;

3) optical noise ambient.

– By mutual agreement, the device containing the infrared module B requires the agreement to its user, for a file transmission.

– Data exchange takes place under the direction of the equipment A, according to the flow set, and no obstacle shall hinder such communication.

– After file transmission, the connection is cut and the session is over.

Beside the link aspects mentioned above, Table 10.3 presents performance parameters, namely:

– The connection delay: time to initialize the connection;

– Latency: the latency of the transmission describes the time of data transmission in the network between two devices;

– The effective throughput is the useful part of the link flow.

Characteristics	Values
Delay	From 250 to 650 ms
Latency (L)	Typically L < 20 ms, but can be under 500 µs
Effective throughput	115 kbit/s link (PDA, mobile phone): data rate 80 kbit/s (70%) 4 Mbit/s link (*laptops* IrDA adapter): data rate 3.5 Mbit/s (88%)

Table 10.3. *IrDA link performance parameters*

The different versions of IrDA standard are listed below:

– The first version of Serial InfraRed (SIR) proposed a flow rate of 9.6–155 Kbps;

– The second version, Advanced Infrared (AIr) (also called Fast InfraRed), was developed around 1998 by the IBM group in Zurich in collaboration with members of the IrDA. The AIr is an enhanced version of SIR that seems to allow multiple devices to connect together, without interference, on a maximum distance of 10 m with a divergence angle of 120°. It offers a rate of 4 Mbps, potentially extended to 16 Mbps. The use of this protocol is typically oriented to a point-to-multipoint (PmP) LAN application with the CSMA/CA protocol, in quasi-diffusion configuration in a limited space (office, living room, car, car of a train, aircraft cabin, etc.). The IBM group's objective was to imperatively ensure the connection even at the detriment of flow rate so that at a certain error rate threshold, a lower bit rate is renegotiated. It does not seem to exist in today's marketed products with all the features; and it also appears that the performances were not up to expectations. However, the vast majority of facilities were proposed for use at 4 Mbps at a distance of 1 m or less, configured in PtP and line of sight (LOS) configuration.

– The third version is called Very Fast InfraRed (VFIR) and was proposed in 1999 by Hewlett-Packard, IBM, and Sharp. The modification occurs primarily on the physical layer and the access protocol link access protocol (LAP). The flow rate obtained is 16 Mbps with a LOS transmission. To achieve this flow rate, a new encoding is used: HHH (1.13) whose acronym includes the initials of researchers who invented it (Hirt Hassner Heise) [HIR 01], but its use seems to provide little improvement in terms of optical budget [FUN 02, GAR 04]. At that time, this new rate of 16 Mbps offered access to new applications such as Ethernet (10 Mbps) or

MPEG video transfer (12 Mbps). But, again, apart from certain products from HP, it does not seem to exist today in any marketed products with all the features.

– Note the Ultra Fast InfraRed (UFIR) is a protocol offering over 100 Mbps (version 1 in 2003), but again, no marketed products.

– IrSimple: around 2005, IrSimple function was to reduce the layers of the IrDA protocol and propose three key stages: connection establishment, data transmission, and interruption of the connection, each time as part of a particular stage and without having to make the LAP, LMP, and OBEX connection protocols. The goal is a higher transmission speed (4–10 times faster). Other benefits for IrSimple IrDA interfaces, especially for mobile applications: a relatively moderate power consumption (200 mW) and a small size. The IrSimple systems are extensively used not only in mobile phones and laptops, but also in the new generation of flat screen televisions on which you can view movies or photos, as well as in printers. Sharp proposed a IrSimple system as a wireless solution.

– Giga Ethernet: the last protocol proposed by the IrDA and previously known as FIR provides a 1 Gbps communication over a few tens of centimeters [KDD 11]. It is always a PtP half-duplex time division. To meet the demands of rapid data transfer, the members of this association were intended to provide a system for early 2010. Examples of use are portable multidevice interface and fast music or film downloading. There is little information about the protocol stacks used and the only elements currently known are the wavelength (1,300 nm) and size (72 mm^3).

10.2.3. *Visible light communication consortium*

Visible light communication consortium (VLCC) is an association that offers Japanese standardization of communication systems, using the LEDs in the visible spectral range. The use of communication is available in personal lighting, offices, the signaling or the architecture lights of a car, the electronic advertising signs, etc. The advantage of this approach is to offer a unique solution for lighting and a low communication data rate (few tens of Kbps to several Mbps). The proposed protocol is a proprietary protocol adapted from Japanese specifications and IrDA PtP data link protocols. The access method seems to be the half-duplex TDMA.

10.3. Point-to-multipoint data link

10.3.1. *IEEE 802.11 IR*

The IEEE 802.11 (WiFi) protocol covers the MAC (access control) and PHY (physical interface) layers of a wireless digital communications system. From the first version of the 802.11 recommendation, there were three proposed protocols. The first one used the frequency hopping spread spectrum (FHSS). The second one was called direct sequence spread spectrum (DSSS), but it was not integrated into the equipment. The last one was based on infrared transmission. The 802.11 IR-PHY standard was developed in 1993 in order to use infrared signals. This IR-PHY part recommendation, published in 1997 and like the DSSS technique, was not used in the equipment but was proposed to the IrDA (FIR).

Such systems operate in LOS or in diffuse mode up to a maximum length of 10–20 m; a wavelength between 850 nm and 950 nm with a pulsed output power limited to 250 mW and an optical security based on IEC 60 825-1 (Class 1); a data rate of 1–2 Mbps and 10 Mbps maximum with a 16 PPM modulation for the 1 Mbps data rate and 4 PPM based on the Gray code for the 2 Mbps data rate.

The subgroup IR-IEEE 802.11 identified the following characteristics for the MAC layer:

– Protocol defined as transparent, a MAC layer (CSMA/CA) for the three PHY layers.

– Each PPDU (physical layer convergence procedure (PLCP) protocol data unit) frame is made (Figure 10.1) of a PLCP header and physical layer service data unit (PSDU) data corresponding to a MAC protocol data unit (MPDU) whose size ranges from 0 to 1,500 bytes (useful data). The PLCP consists of a preamble for synchronization (SYNC – from 57 to 73 bits or pulses and the start frame (start frame delimiter – SFD) has 4 bits – 1001 value). It also consists of a header of 67 bits:

- The first field "Data Rate" identifies the proposed data rate. It is initially limited to two values (000 for 1 Mbps and 001 for 2 Mbps), but it offers eight possibilities.

- The second field "DC Level Adjustment" is a predefined sequence of 32 bits, depending on the data rate and allowing the receiver to adjust the level and identify the DC component of the received signal.

- The third field, called "Length" and coded on 16 bits, indicates the length in bytes of the frame. Associated with the field "Data Rate", this information allows the receiver to estimate the end of the frame.

- The fourth field "Frame Check Sequence" is a detecting and an error decoder algorithm (CCITT CRC-16) coded on 16 bits.

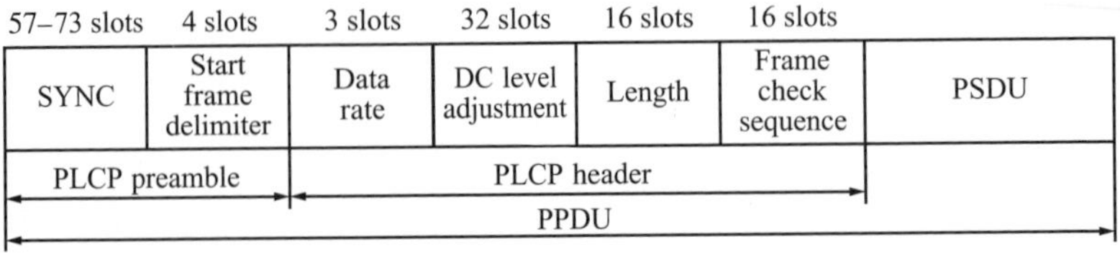

57–73 slots	4 slots	3 slots	32 slots	16 slots	16 slots	
SYNC	Start frame delimiter	Data rate	DC level adjustment	Length	Frame check sequence	PSDU
PLCP preamble		PLCP header				
PPDU						

Figure 10.1. *802.11 IR frame*

10.3.2. *ICSA – STB50 (IEEE 802.3 – Ethernet)*

ICSA protocol [ICS 11] is a proposed standard by a Japanese association. It is an extension of the ARIB STB-T50 standard to 1 Gbps. This standard defined the communication protocol of a multiaccess LAN infrared system at 10 Mbps. In addition to the strategic issues in relation to the IrDA and marketing protocol related to the Japanese market specificities, this standard is intended to adopt, in the wireless optical domain, a compatibility with the 2002 ISO/IEC 802.3 Ethernet Recommendation.

The security aspect is based on IEC 60825-1 with a Class 1 product aim. This standard operates for both exterior and interior systems, with distances of 1–10 m, regardless of the wavelength. Based on this ICSA protocol, 10 Mbps and 100 Mbps VIPSLAN products were the first optical communication systems to offer a wireless optical communications network; it was only sold in Japan by JVC in the mid-2000s and in direct competition with WiFi solutions. The first version of WiFi products has pushed this device in the background.

Other solutions [MRV 11, SCH 06] have been implemented by making a direct electrical/optical conversion and vice versa in PtP Ethernet baseband.

This protocol is based on the IEEE 802.3 standard (Ethernet). Proposed on the CSMA/CD access method, it details a number of characteristics depending on the data rate under consideration:

– Ethernet 10 Base T full-half-duplex (IEEE 802.3):

  - Interframe time: 9.6 µs;

  - Minimum length of frame: 64 bytes;

  - Maximum length of frame: 1,518 bytes;

  - Preamble: 6.4 µs (8 bytes);

  - Maximum throughput: 9.87 Mbit/s (812.74 frames/s × 12,144 bits);

- Manchester code baseband transmission;

- Modulation at two levels;

– Fast Ethernet 100 Base FX full/half-duplex (IEEE 802.3u and 802.3):

- Interframe time: 0.96 µs;

- Preamble, similar minimum and maximum length of frame;

- Maximum throughput: 98.7 Mbit/s (127.4 8 frames/s × 12,144 bits);

- Transmission baseband coding 4B/5B and NRZI;

- Modulation at two levels;

– Gigabit Ethernet 1,000 Base X full-duplex:

- Interframe time: 0.096 µs;

- Preamble, similar minimum and maximum length of frame;

- Minimum length of frame: 64 bytes;

- Maximum length of frame: 1,518 bytes (or 9,018 bytes – Jumbo);

- Transmission baseband NRZ coding and 8B/10;

- Modulation at two levels.

10.3.3. *IEEE 802.15.3*

The IEEE organization and the IEEE 802.15.3 committee created a working group called 3C (3C or TG3c Task Group) in March 2005. The TG3c (Millimeter Wave WPAN Alternative PHY) developed an alternative physical layer based on millimeter wave for an existing wireless personal area network (WPAN), 802.15.3 dating from 2003. The finalization of this standard offers data rates of several Gbps.

The propagation characteristics at 60 GHz could be considered similar to the wireless optical and the MAC layer processing present similarities. There is not yet a multiaccess management but works are performed for an application of this type with a TDMA access.

10.3.4. *IEEE 802.15.7*

In the field of wireless optical communication using visible light and following the studies of the VLCC, Japanese association papers were available from 2007 and a new IEEE 802.15 working group [WON 07] was created (IEEE 802.15.7) to propose PtP and PmP solutions. The main features are:

– PtP and PmP access in line of sight;

– centralized or distributed configuration with mobility;

– carrier sense multiple access with collision avoidance (CSMA/CA);

– acknowledgment mechanism for reliable transfer;

– color quality indication (CQI);

– network access managed by coordinator equipment.

10.3.5. *Optical wireless media access control*

The optical wireless media access control (OWMAC) protocol [OME 09] was proposed as part of a European project called Omega in 2009 in order to obtain a MAC layer suitable for wireless broadband optical communications regardless of the wavelength. The development of this protocol was made from the existing MAC protocols and new proposals.

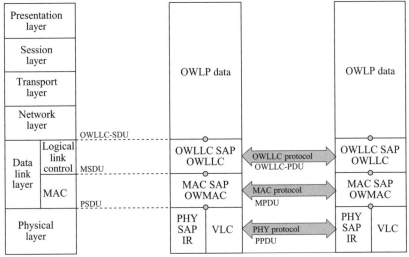

Figure 10.2. *OWMAC layer*

As shown in Figure 10.2, the OWMAC sublayer corresponds to the MAC sublayer of the ISO/IEC standard of the OSI reference model. This is an MPDU that messages between a PSDU and a MAC service data unit (MSDU). This OWMAC service is provided through the service access point (MAC SAP) to the LLC sublayer.

The LLC sublayer controls the packets exchange. It is called optical wireless logical link control (OWLLC). This service is provided through the SAP OWLLC to the upper-layer protocol.

The OWMAC protocol requires the following physical configuration: PLCP header and PPDU. Figure 10.3 shows the structure of the physical environment.

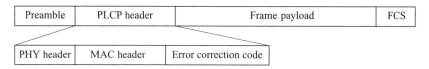

Figure 10.3. *OWMAC frame*

The preamble or synchronization sequence is used for clock recovery (line clock recovery – LCR). The PLCP header contains the MAC and PHY headers with an error correction code (ECC). The ECC helps to protect the acknowledgment frames. The useful frame (frame payload), followed by the frame check sequence (FCS), closes the physical structure.

The purpose of the OWMAC architecture is to be fully distributed. No equipment acts as a central coordinator. This is achieved by the exchange of beacon frames. The periodic transmission of this tag allows the device detection, performs a dynamic organization of the network, and provides a mobility. These tags provide the basic temporal structure of the network and carry out the reservation planning for the medium access. The underlying mechanisms are TDMA access and multisector transmission.

This protocol provides a data rate adaptation from 128 to 1,024 Mbps with a half-duplex or full-duplex communication. To meet the imposed constraints of many services and levels of security, the protocol is also able to manage a mesh or a star topology. It also includes functions such as an urgency message, even in the case of traffic congestion and QoS parameters.

This protocol, free download [OME 10a], was implemented in two prototypes: the first one in visible light has a 100 Mbps data rate with a coverage of about 12 m^2

and the second one in infrared offers a throughput of 300 Mbps with a coverage of about 30 m^2. The next step would be an industrialization process.

10.4. Summary

The proposed data link protocols in wireless optical communication field are or have been numerous and varied. To each proposal, the progress in optical analogic and digital modules was included to increase the coverage or the data rate or both simultaneously.

Several data link layer protocols currently meet all possible configurations from PtP communication to a network communication PmP and solutions are proposed for wavelengths in the visible or infrared light (Table 10.4).

	Visible	Infrared
Point-to-point	VLCC/IEEE 802.15.7	IrDA/IEEE 802.15.3
Point-to-multipoint	IEEE 802.15.7/OWMAC	OWMAC

Table 10.4. *Wireless optical link protocols*

Chapter 11

Installation and System Engineering

"Innovation is an alliance between research,
marketing, instinct, imagination, product and industrial courage".
Antoine Riboud
(1918–2002)

11.1. Free-space optic system engineering and installation

11.1.1. *Principle of operation*

Basically, the principle of transmission of a free-space optic (FSO) is based on the transmission of a divergent beam in line of sight (LOS).

In Figure 11.1, the equipment is placed along the top of each building and they are pointed toward each other. The A equipment of the X site sends digital information to the Y site via a modulated optical beam, such as a laser diode. This beam is voluntarily slightly divergent to avoid problems of misalignment. Part of the wave front is collected on the receiving optics antenna and focused on the receiver diode (e.g. PIN diode) of the B equipment placed on the Y site. This operation is performed in the opposite direction, from Y to X (full-duplex).

Each device consists of several modules:

– For the transmission function:

- the connection interface: electrical or optical to send and receive digital data;

- the electrical/optical conversion module (case of optical interface);

- filtering and amplification of the electrical digital signal;

- the emission optical module containing the laser.

– For the receiving function:

- the receiver optical module containing the diode;

- filtering and amplification of the electrical digital signal;

- the electrical/optical conversion module (case of optical interface);

- the connection interface: electrical or optical to send and receive digital data.

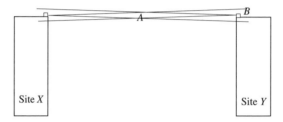

Figure 11.1. *FSO between X and Y sites*

In most cases, a supervision software is supplied with the equipment. This software allows us to configure the connection and obtain qualitative and quantitative information of the different modules.

According to the manufacturers, additional features are implemented:

– a tracking system,

– a limited data rate radio link for an emergency in case of laser cut-off,

– etc.

An example of FSO equipment structure is shown in Figure 11.2.

11.1.2. *Characteristics*

To avoid a too detailed description of the various products, a presentation of the general parameters is given below.

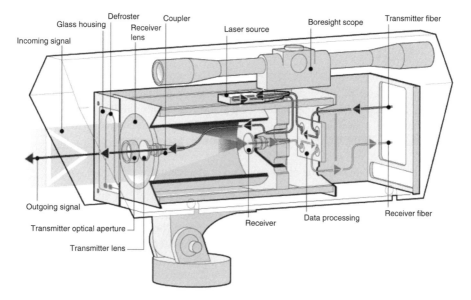

Figure 11.2. *FSO equipment example [BOU 04]*

We will restrict ourselves to the products that have the following properties:

– laser technology whatever the wavelength, power, and category;

– a numerical data transport function;

– a point-to-point connection or system, LOS;

– transmission rates from megabits per second to several terabits per second.

11.1.2.1. *Principal parameters*

The principal parameters that must be taken into account in the definition of optical links are range, safety, and data rate.

11.1.2.1.1. Range

According to the equipment, it varies from a few tens of meters to several kilometers. Certain manufacturers give a maximum range, others specify the typical range for various weather conditions, and others propose a "recommended" range, integrating a margin around a maximum value. These values must be taken as orders of magnitude, and not as absolute values.

It should also be noted that the calculation of the margin of a given link gives more indication about the link quality of service (QoS).

11.1.2.1.2. Safety

The class laser of the system equipment is an important factor of consideration because this determines more or less easily the modalities for the installation and maintenance of a laser link. The parameters to be taken into account when defining laser category are the signal wavelength, the power, and the beam form.

11.1.2.1.3. Data rate and type of recommended application

Many systems are transparent to data rate and to protocol, for an often relatively important data rate range. The applications then depend on the maximum capacity that the system can transmit, and invariably occur in the telecommunications or in data-processing world. For instance, a transparent system up to 200 Mbps will be able to transmit STM-1, FDDI, Fast Ethernet, or GigaEthernet signals.

11.1.2.2. *Secondary parameters*

Other parameters should also be taken into account for the choice of a system such as:

– the wavelength at which the optical link operates: this parameter influences the link margin and, by consequence, the QoS;

– the type and number of optical transmitters: this also influences the link margin;

– the alignment control (tracking): it potentially offers better protection against shock and vibration;

– additional safety components: an automatic beam cut-off in case of presence detection, for example, or a security key to access equipment;

– a simple process of implementation and maintenance;

– simple, convivial supervision software allowing the management of two (or more) elements of links from only one site;

– the cost of the system.

11.1.2.3. *Examples of FSO systems*

More than several hundred systems were listed in 2010, so it is difficult to represent these products by a single image. For more information, the reader should refer to the manufacturers' sites indicated in Table 11.1.

Names	City	Country	Website
AirLinx	Boston	USA	http://www.airlinx.com
Aoptix	Campbell	USA	http://www.aoptix.com
CableFree	Hampton Hill	UK	http://www.cablefree.co.uk
Communication By Light	Munster	Germany	http://www.cbl.de
Fsona	Richmond	Canada	http://www.fsona.com
Katharsis	St Petersburg	Russia	http:/www.optica.ru
Lase	Wesel	Germany	http://www.lase.de
LightPointe	San Diego	USA	http://www.lightpointe.com
LSA	Exton	USA	http://www.lsainc.com
MRV	Yokneam	Israel	http://www.mrv.com
Optel	Hamburg	Germany	http://www.optel.de
PAV Data Systems	Cumbria	UK	http://www.pavdata.com
Plaintree Systems	Arnprior	Canada	http://www.plaintree.com
RedLine	Kyalami	South Africa	http://www.redlinesa.com

Table 11.1. *Some FSO manufacturers*

11.1.3. *Implementation recommendations*

In general, FSO equipment is set up as a radio-relay system:

– installation on a high point (such as a building, a pylon, and a water tower);

– LOS, without present or future obstacles;

– installation time less than 1 day for a link.

However, due to the technology used and in addition to the safety requirements related to the laser class equipment, some elements have to be taken into account at the time of installation (see Figure 11.3).

Very precise alignment is necessary, given the characteristics of the equipment (i.e. low divergence of the laser beam). The coupling of the optical link is characterized by the alignment of the transmitter and the receiver. These can be disturbed following mechanical vibrations. The fitter of the communication system must:

– fix the materials on a rigid support or a load-bearing wall so that it is subjected to less possible vibration or shock (for instance, edges of walls and sides of walls);

– avoid the direct optical alignment from the Sun's rays;

– avoid the proximity of elements that can cause atmospheric turbulence (such as chimneys and reflective surfaces).

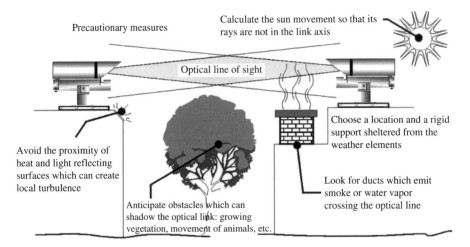

Figure 11.3. *FSO implementation: potential problems to avoid*

11.1.4. *Optic link budget*

One of the important elements to know in free-space optical transmissions is the margin of the laser link. In fact, like radio operator equipment or radio-relay systems, it is of primary importance to know the margin of a given link. When a link is installed, mathematical models allow us to calculate the availability of the link for one year, or for the most unfavorable month.

The first step consists of calculating the link margin. This element allows us to know the capacity of the laser equipment to transmit numerical data in spite of the variations in weather conditions.

To use the prediction models, the necessary parameters of the equipment are:

– the emitted power;

– the receiver sensitivity;

– the receiver capture area;

– the divergence of the emitted beam.

From these data, we can determine the value of the link geometrical attenuation, then the laser link margin, and thus its availability.

11.1.4.1. *Geometrical attenuation concept*

The beam emitted by the transmitter being divergent (Figure 6.1), the receiving cell will collect only a fraction of the energy emitted:

$$Aff_{geometric} = \frac{S_d}{S_{capture}} = \frac{\frac{\pi}{4}(d\theta)^2}{S_{capture}}$$

where

– θ is the divergence of the beam (e.g. 1.3 mrad).

– d is the transmitter – receiver distance.

– $S_{capture}$ is the capture area of the receiver (e.g. 0.005 and 0.025 m^2).

– S_d is the area of the beam at distance D.

Equation 11.1. *Geometric attenuation*

The attenuation is given in decibels by $Aff_{dB} = 10\log_{10}(Aff)$.

The geometrical attenuation is thus a function of the beam divergence, the distance, and the receiver capture area. Table 11.2 gives some examples of geometrical attenuation values.

Divergence (mrad)	Distance (km)	Geometrical attenuation (dB)
1	5	29
1	8	33
3	2	37

Table 11.2. *Values of geometrical attenuation for various distances*

11.1.4.2. *Link margin concept*

After calculation of the geometrical attenuation of a laser link, the margin is determined as follows.

Emission power (P_e), receiver sensitivity (S_r), and data given by manufacturers allow the calculation of the link margin using the following formula [11.2]:

$$M_{link} = P_e + |S_r| - Aff_{Geo} - P_{syst}$$

where

– M_{link} is the link margin (dB).

– P_e is the power of the emission signal (dBm).

– S_r is the receiver sensitivity (dBm).

– Aff_{Geo} is the geometrical attenuation of the link (dB).

– P_{syst} is the equipment loss (dB).

Equation 11.2. *FSO margin*

In Table 11.3, some examples for the calculation of a link margin of 500 m for three industrial equipments are presented.

These link margins are the basic elements to then understand the laser signal impairments due to climatic phenomena mentioned in Chapter 5 (fog, mist, rain, snow, scintillation, etc.).

	Equipment A	Equipment B	Equipment C
Emission power	8 dBm	10 dBm	17 dBm
Sensitivity	−38.2 dBm	−36 dBm	−32 dBm
Geometrical attenuation (D = 500 m)	(θ = 1 mrad; Capture area: 0.002 m²) 15.9 dB	(θ = 2 mrad; Capture area: 0.005 m²) 22 dB	(θ = 1 mrad; Capture area: 0.005 m²) 15.9 dB
System losses	3 dB	2 dB	2 dB
Link margin	10.8 dB	20.8 dB	29.8 dB

Table 11.3. *Link margin of three typical systems for a distance of 500 m*

11.1.5. *FSO link availability*

11.1.5.1. *Characteristics*

This section presents examples of available research for a given link with the characteristics considering the two latest FSO equipments from Table 11.3. The availability and reliability of an optical link are highly dependent on proposed equipment characteristics as well as climatic and atmospheric parameters of the selected site such as rain, snow, and fog. Through the algorithms it implements, an "FSO Prediction" tool integrates the different physical phenomena potentially responsible for the long blackout. This tool consists of a complete graphical user interface (GUI), meteorological data from French and English stations over a decade, and computing code module. It assesses the QoS of a free-space optical link in terms of probability of connection and disconnection times.

In our example, we considered the following parameters:

– Link distance: 500 m;

– Manufacturer: there are three products A, B, and C from different origins (product A has the same margin like that of product C);

– Equipment: providing a link at 1 Gbps with Ethernet fiber interface;

– Model: we apply the models attenuation introduced in "FSO Prediction" software (see Chapter 5) giving the attenuation values due to aerosols (fog), rain, snow, and scintillation;

– Sites: three sites were studied (Paris, Rennes, and Sainte-Maxime).

The QoS and availability search is a three-step process: the first step consists of puting into the computer the site and equipment characteristics; the second step is the calculation process; and the third step is the analyzing of the results.

In the first step, we define the sites and facilities characteristics:

– site name, its altitude, latitude, longitude, height, and orientation of the equipment; then the environment with moisture, *albedo*, roughness, etc.;

– equipment: wavelength, data rate, power, sensitivity, divergence, number of transmitters, etc.

From the sites and equipment data, the software calculates various elements such as:

– the distance and the average height, the link margin, the sensitivity to misalignment, etc.;

– the attenuation of the rain, snow, fog, and scintillation;

– the percentage of occurrence of values greater than the margin at the respective sites.

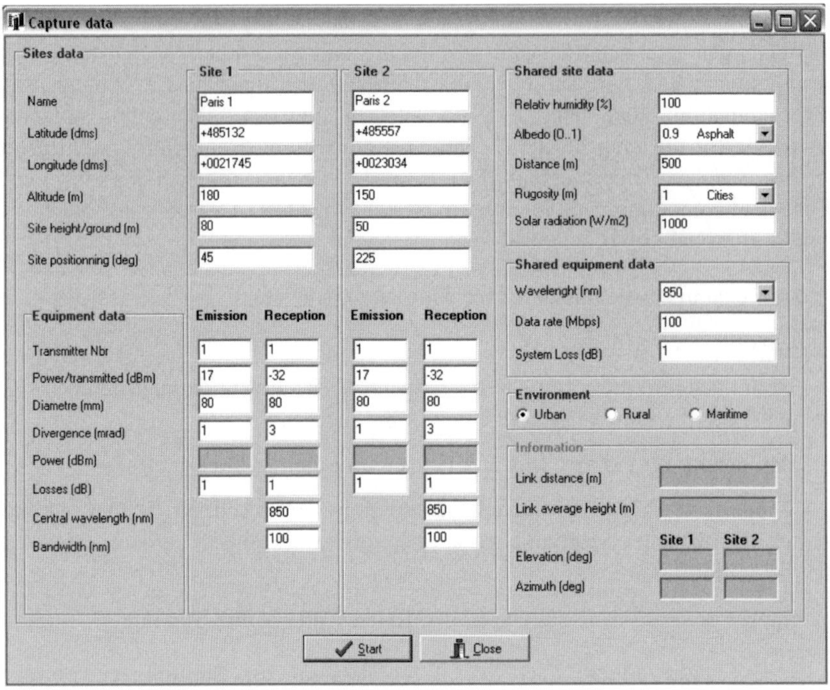

Figure 11.4. *Data site and equipment example for Paris*

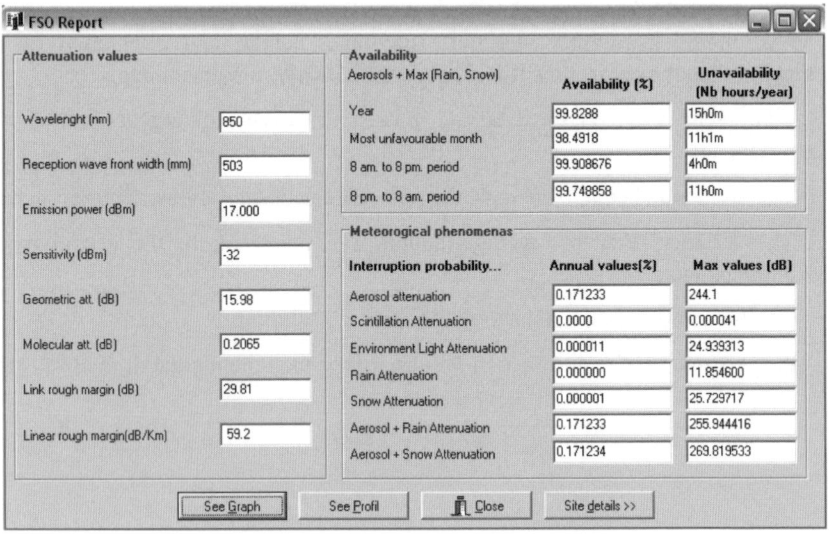

Figure 11.5. *Example of results for Paris*

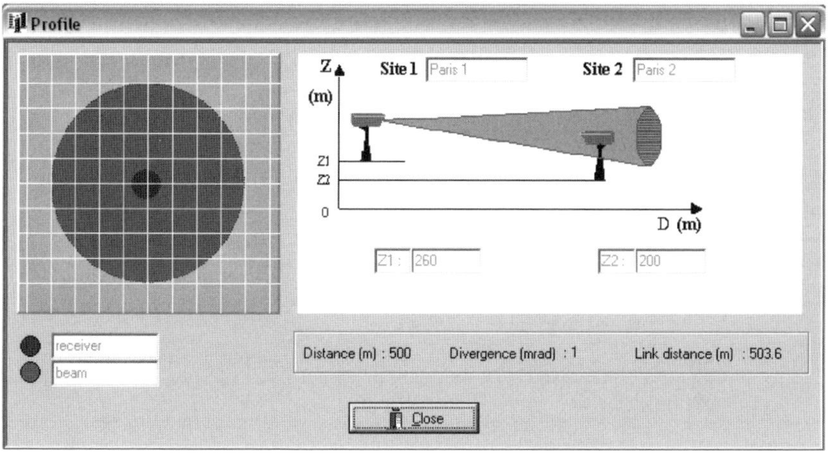

Figure 11.6. *Example of equipment profile for Paris*

The calculation result is displayed as a table and chart to know:

– the link margin;

– the availability throughout the year, the worst month, and the day and night time;

– the attenuation values of climatic phenomena and their percentage of appearances on the considered site;

– the profile of the link to so as know the sensitivity of the equipment to a misalignment.

11.1.5.2. *Results*

The following tables show the margins and the availability values for each equipment and for each site. The availability of the optical link is proposed as a percentage per year.

From these examples, we derive two important points:

– The availability of an optical link is very dependent on local climatic configuration (fog mainly).

– The management of availability and intrinsic qualities of selected equipment (link margin) is also dependent on the link distance (at identical climatic conditions).

Site: Paris

	Equipment A	Equipment B	Equipment C
Margin (dB)	10.8	20.8	29.8
Specific margin (dB/km)	21.6	41.6	59.6
Worst climatic phenomenon (dB)	Fog (169)	Fog (169)	Fog (169)
Link availability (%)			
Year	99.72	99.92	99.94
Worst month	98.90	99.59	99.59
Day time: 8 am–8 pm	99.77	99.89	99.91
Night time: 8 pm–8 am	99.68	99.95	99.98

Site: Rennes

	Equipment A	Equipment B	Equipment C
Margin (dB)	10.8	20.8	29.8
Specific margin (dB/km)	21.6	41.6	59.6
Worst climatic phenomenon (dB)	Fog (342)	Fog (342)	Fog (342)
Link availability (%)			
Year	97.86	98.45	99.03
Worst month	91.23	92.87	95.21
Day time: 8 am–8 pm	98.54	98.83	99.25
Night time: 8 pm–8 am	97.19	98.06	98.81

Site: Sainte-Maxime

	Equipment A	Equipment B	Equipment C
Margin (dB)	10.8	20.8	29.8
Specific margin (dB/km)	21.6	41,6	59.6
Worst climatic phenomenon (dB)	Fog (173)	Fog (173)	Fog (173)
Link availability (%)			
Year	99.16	99.75	99.86
Worst month	98.36	98.49	98.77
Day time: 8 am–8 pm	99.81	99.84	99.89
Night time: 8 pm–8 am	99.40	99.66	99.84

Table 11.4. *Availability results of selected links*

11.1.6. *Summary*

In this section are listed various potential applications of free-space optical links.

The first results of studies suggest that this technology could answer to the famous "last mile" syndrome, that is to say the ability to connect directly in broadband, in point-to-point, a limited geographical area.

A result of American study indicates that more than 95% of the business areas are within a mile of a fiber optic node, indicating an important potential of high-speed connections for the use of FSO.

It is worth mentioning that in the example of a network configuration, a mixed solution combining free-space optical links and microwave link would be the most suitable because availability will become permanent regardless of distance or weather conditions. This new technology is not in competition with existing technologies, but complementary.

For some geographical specificities (classified or private risk sites), the free-space optical link has some advantages that can provide technology based on radio or the optical fiber.

Its use can also be proposed as part of a connection for temporary action events, private intersite, emergency, or to close a fiber-optic loops.

11.2. Wireless optical system installation engineering in limited space

This section is dedicated, in the first instance, to an example of installation of a wireless optical system in a limited space and, then in a second instance, to the technical characteristics of such a device.

11.2.1. *Habitat structure*

First of all, it is important to know the structure of the domestic "housing". This information is used to define the volume characteristics of each part of the habitat.

A statistical study of housing in France [TEC 08] is presented in the paragraph below. It is important to note that the sample is not representative of the French population: it is a subjective and biased sample, limiting the number of respondents to a few hundred, with occupational category limited to senior management.

But this approach, supported by similar European data [OME 08a] and [OME 08b] shows nevertheless, relevant qualitative information on the habitat.

Table 11.5 shows more accurately the distribution of areas of different parts of a habitat. The last column indicates successively the mean, median, and standard deviation.

Surface (m²)	Ground floor	0–19	20–29	30–39	40–49	+50	M/median/σ
Living room (%)	97	8	28	32	20	12	33.2/30/12.3

Surface (m²)	Ground floor	0–10	10–19	20–29	+30		M/median/σ
Kitchen (%)	97	27	61	10	2		12.4/12/5.76
Bedroom 1 (%)	60	10	80	10	0		13.2/12/4.12
Bedroom 2 (%)	30	10	80	10	0		13.2/12/4.12
Bedroom 3 (%)	20	12	78	10	0		13.5/12/5.58
Bedroom 4 (%)	12	12	72	16	0		13.8/12/4.91
Office (%)	50	24	64	12	0		12.43/11/5.62
Corridor RDC (%)		68	32	0	0		7.7/6/4.89
Corridor 1st floor (%)		82	18	0	0		5.7/5/4.36

Table 11.5. *Distribution of room surfaces*

11.2.2. *Statistical analysis and coverage area*

The first analysis on the structure of the habitat is linked to a connectivity hierarchy. From the access point located in the living/dining room, the connectivity hierarchy is generally room 1, often the parental bedroom and bedroom 2 and/or office and then the other rooms (kitchen, garage, attic, etc.).

In the case of a wireless optical system, the optical beam does not pass through walls. This physical limit provides the implementation of a simple and intuitive architecture, without disruption in the other rooms or the next-door apartment.

The proposed example is a duplex apartment consisting of a concrete interlevel, a dense metallic mesh, and a WiFi radio base station: (a) ground floor and (b) first floor (light gray, excellent reception; white, medium reception; and dark gray, limited or no reception) (Figure 11.7), and an installation with only wireless optical devices in limited space (Figure 11.8).

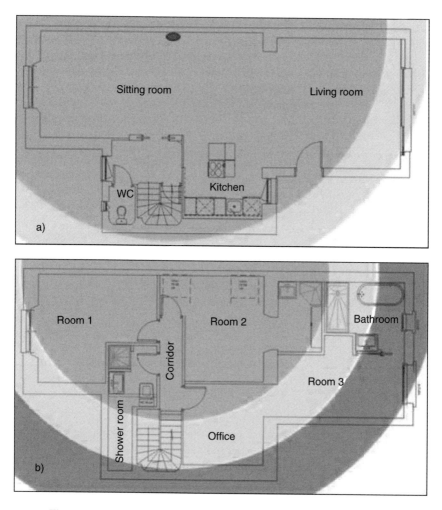

Figure 11.7. *Example of architecture of a wireless radio system like WiFi*

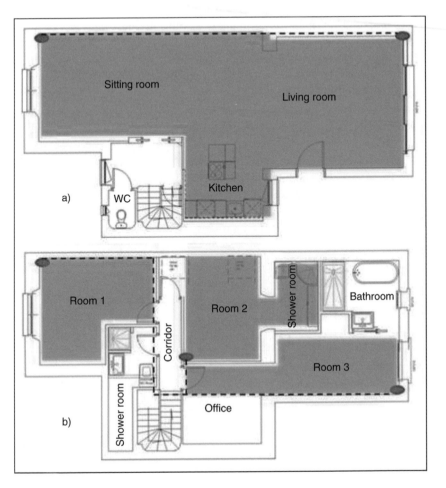

Figure 11.8. *Example of architecture of a wireless optical system*

In the architecture of a wireless optical system, there is at least one wireless optical transmission/reception device per room, called the base station (gray ellipse). Each base station communicates with terminals present in every room via a wireless optical communication. Finally, these terminals are connected or integrated to multimedia communications equipment (PC, monitor, PDA, etc.).

The link with other rooms (dashed lines) occurs with an inter-room connectivity of electric (Ethernet cable, power line, etc.) or optical type (plastic fiber, single-mode silica fiber, etc.). Depending on the media and signal processing used, the connectivity data rates range from several tens of megabits per second to several gigabits per second. It is possible of course to implement other architectures such as a star network, for example. This network is connected to the outside network through an access point (network access point – AP) that can be an electrical (xDSL) or an optical (FTTx) connection.

So as a first step, a wireless optical device in limited space can be implemented in two basic rooms:

– living/dining room;

– … room.

Next, Table 11.6 shows the cumulative statistical distribution of surfaces; it defines the percentage of concerned population compared with a covered communication area (Gaussian distribution). For example, 84% of the population of the sample have a living/dining room surface less than or equal to 45.5 m^2.

Population (%)	84	97.5	99.85
Living room surface up to (m^2)	45.5	57.8	70.1

Population (%)	84	97.5	99.85
Bedroom surface up to (m^2)	17.3	21.4	25.6

Table 11.6. *Cumulative distribution of surfaces*

Note that there is generally a "multimedia zone" in the living/dining room (screen, tuner, DVD player, etc.) and it corresponds to at least 100% of the coverage area of a room (approximately 25 m^2).

From this information, it is possible to determine a reference example of a room to calculate the optical margin necessary for a given coverage in all possible configurations cases.

This reference example has a dimension of 5 m × 5 m or 25 m^2 and 2.50 m high. From this reference, the most pessimistic location cases for wireless optical system are presented below (Figure 11.9):

– Base station (letters):

  - A: located in the middle of the ceiling;

- B: located above the door (2.20 m high);

- C: located on a socket, e.g. telephone, Ethernet, PLC (0.20 m high);

– Terminal (numbers):

- 1: located in the lower opposite corner of the room (0.20 m high);

- 2: at a height equivalent to the top of a door, e.g. speaker, motion sensor (2.20 m high);

- 3: installed on the floor in the middle of the room.

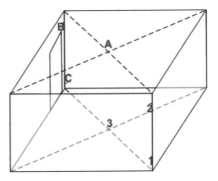

Figure 11.9. *Simulation in a reference room*

11.2.3. *Optical link budget*

The simulation was performed with classical optical devices in a 25 m^2 reference room. A third of the population has a living/dining room area corresponding to that surface. It also corresponds to 100% coverage of a multimedia area and almost 100% coverage of a room. To calculate the optical link budget in a limited space (see Chapter 6), the goal is a complete coverage and a calculation of the maximum values of angles and distances has previously been made. These target values are given in Table 11.7 (diffusion – DIF and line of sight – LOS).

	Maximum angle (°)			Maximum distance (m)		
	1	2	3	1	2	3
A	57	85	0	4.2	3.6	2.5
B	74	90	57	7.4	7.1	4.2
C	56	72	36	8.4	7.4	5.8
	DIF	LOS	DIF	DIF	LOS	DIF

Table 11.7. *Maximum angle and distance*

From these data, it is possible to define the maximum angular and distance characteristics of a wireless optical device in a limited space. Following these three configurations, these characteristics are shown in Table 11.8.

Case	FOV or DIV (°)	Maximum distance (m)
A	±85	4.2
B	±45	7.4
C (DIF)	±36	8.4

Table 11.8. *Objective of system characteristics*

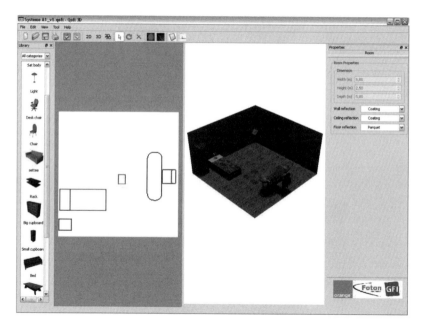

Figure 11.10. *Example of a furnished room*

Figures 11.10–11.13 show an example of a reference furnished room and the three configurations. They are made with a free download software application called QOFI [QOF 11], which allows us to build a specific environment from the GUI and to simulate a connection according to the wireless optical equipment characteristics, walls and furniture materials.

Table 11.9 shows the attenuation values of these three geometric configurations with the most pessimistic location of the terminal and the following systems values ($P_e = 30$ dBm, $S_e = -40$ dBm, and $A_{eff} = 10$ mm^2).

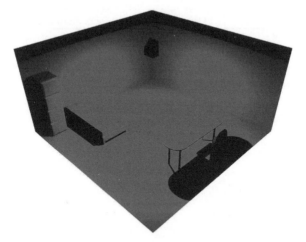

Figure 11.11. *"A" configuration example*

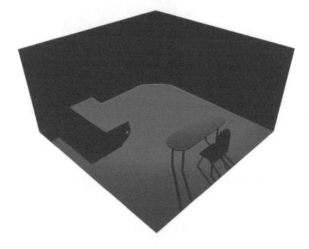

Figure 11.12. *"B" configuration example*

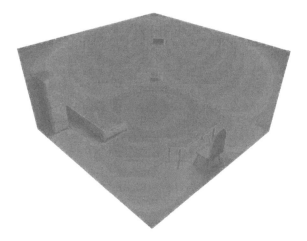

Figure 11.13. *"C" configuration example*

Configuration	Geometric attenuation (dB)
A	68
B	71
C (DIF)	124

Table 11.9. *Calculated geometric attenuation*

Configuration "A" provides the smallest maximum distance but to the detriment of a very important DIF and field of view (FOV) couple values. To avoid a large FOV, one solution is to place several systems equidistributed in the ceiling. Thus, at the same distance, it is possible to significantly reduce the DIV and FOV couple values. This approach is one of the axes of wireless optical system solution in the visible light. Indeed, each lighting plate is also a wireless optical communication device.

Configuration "B" is a good compromise between the maximal distance and a limited DIV and FOV angular couple. In addition, its installation can also be performed at the time of a home renovation.

Configuration "C" is the easiest to install but also the most complicated to implement because the diffuse propagation causes a very important geometrical attenuation. The optical link budget of actual systems does not allow this configuration, except in the case of a spatial distribution of several wireless optical devices or a MIMO approach.

11.2.4. *Optimization of indoor wireless optical system*

The geometric attenuation values in Table 11.9 are significant and the optical link budget of an indoor wireless optical system must be higher to achieve a positive margin and correct operation.

According to recently produced demonstrators and those expressed below, several approaches are considered:

– Omega demonstrator in visible light (VLC) [OME 11b]:

   - broadcast at 100 Mbps;

   - 10 m^2 coverage at 2 m;

   - 16 LED TX (transmitting) inserted in the ceiling;

   - 12 MHz bandwidth with DMT signal processing (digital multitone);

– Demonstrator of Keio University (Japan) in visible light (VLC) [SPA 09]:

   - half-duplex at 100 Mbps;

   - 1 m^2 coverage at 2 m;

   - 20 LED TX and an RX receiver for the base station;

   - 2 LED TX and an RX receiver with an automatic pointing (mirror) for the terminal;

– Techimage demonstrator in infrared (IRC) [TEC 11]:

   - half-duplex at 1 Gbps;

   - 0.01 m^2 coverage at 0.5 m;

   - 1 TX / RX;

   - bandwidth of 1.25 GHz modulated OOK (on–off keying);

– Omega demonstrator in infrared (IRC) [OME 10a]:

   - half-duplex at 1 Gbps;

   - 1.5 m^2 coverage at 3 m;

   - MIMO with 3 TX/RX;

   - 1.25 GHz bandwidth signal processing with OOK;

– Omega demonstrator in infrared (IRC) [OME 11b]:

   - half-duplex at 300 Mbps;

- 30 m^2 coverage at 4 m;

- MIMO with 7 TX/RX;

- bandwidth to about 300 GHz signal processing with OOK.

To achieve the throughput of 1 Gbps, the first thing to consider is the realization of an optronic module suitable to wireless optics. The majority of available modules are configured for use in optical fiber.

At the emission, the security parameter is essential and the optical part can be equipped with an economical material diffuser (PCV, for example). The required optical power is achieved either by combining several LEDs or laser diodes with a possible spatial management of the emitted beams (MIMO) [HAY 10] or by using a high-power laser (Chapter 8).

At the reception, the actual sensitivity of PIN or APD photodiodes is close to the theoretical value and the search for additional gain is not currently economically viable. The matrix approach with integration of the amplifier, like CCD cameras, seems most appropriate. Spatial management of the received beams (MIMO) is also a better option. Finally, it is possible to obtain an optical gain by the use of focusing devices more effective than the aspherical hemisphere, for example, the fisheye optical approach (Chapter 8).

In the field of signal processing, OOK modulation is the simplest and easiest to implement. Treatments such as OFDM and equalization allow a reduction in bandwidth and therefore a possible gain in the optical budget, compared with the lower signal-to-noise ratio (SNR). These treatments offer, in all cases, a better resistance to intersymbol interference (Chapter 9). Serious study and prototyping should define the technical and economic balance of such a solution, such as the cost of the number of bits per channel and the damage on the SNR.

Finally, at the network layer level, the most suitable current protocol for high data rate, such as the gigabit per second, is the one developed in the Omega European project; the mesh network TDMA approach can maintain a good efficiency of the available bandwidth without a coordinator management (Chapter 10). But some new protocols from the radio domain, like IEEE 802.11 ad, could also be used for wireless optical devices. In addition, the recent development of specification allowed us to incorporate QoS parameters and to optimize the energy resources needs.

Chapter 12

Conclusion

"...But since I designed to employ my whole life in the search after so necessary a science, and since I had fallen in with a path which seems to me such, that if any one follow it he must inevitably reach the end desired, unless he be hindered either by the shortness of life or the want of experiments, I judged that there could be no more effectual provision against these two impediments than if I were faithfully to communicate to the public all the little I might myself have found, and incite men of superior genius to strive to proceed farther, by contributing, each according to his inclination and ability, to the experiments which it would be necessary to make, and also by informing the public of all they might discover, so that, by the last beginning where those before them had left off, and thus connecting the lives and labours of many, we might collectively proceed much farther than each by himself could do".

René Descartes (1596–1650)
Discourse on the method,
Chapter 6, 1637

For many years, ergonomists, researchers, and engineers have dreamt of offering the gift of ubiquity. Beyond the philosophical and psychological aspects, this can result in "wireless" connectivity everywhere for everyone, regardless of location or time.

This "wireless" connectivity can be improved by the integration of intermodality, such as augmented reality and virtual reality, requiring a large bandwidth or throughput with a significant bilateral precise tracking device.

Technical and commercial advances abound in this sense, with the success of more than two billion people communicating via cell phones and hundreds of millions of people every day using WiFi hotspots.

The conventional radio solutions (300 kHz–3,000 GHz) have so far met the expectations of customers. But due to increasing demand in throughput, spectral congestion becomes critical, not only in relation to the powers implemented and electricity consumption, but also in relation to the questions they raise for some users' radio systems concerning the safety aspect [IAR 11].

An alternative is proposed by wireless optical technology using the infrared or visible spectral range. Its use has been developed in many areas of telecommunications: the remote control, the visible light communication (VLC), the free-space optical (FSO) links, intersatellite links, and indoor wireless optical links.

From a regulatory point of view, their implementation does not require any frequency authorization or any cost to get a license.

A techno-economic example, already commonly validated, is the infrared remote control.

Another validated techno-economic segment is the FSO links, which are an alternative to microwave links and to optical fiber cables to meet the growing needs of broadband telecommunications.

This solution uses low-power laser beams to ensure a negligible impact on the environment. The main advantages over optical fibers are their low cost, flexibility, installation, and deployment speed. No civil engineering works are required.

Availability of components (lasers, receivers, modulators, etc.) widely used in optical fiber telecommunications greatly reduces the cost of such equipment. Manufacturers today offer relevant systems whose interfaces match the industry standards (Gigabit Ethernet, for example) allowing the interconnection of local area network to remote sites.

Another contemporary application relates to intersatellite links for which low power consumption and small size favored the optical wireless solution instead of embedded radio systems.

The indoor wireless optical systems are entering a phase of commercial maturity in the visible spectrum (VLC) with applications such as broadcast information on mobile phones and data traffic in vehicles (ITS).

In the infrared spectrum, we are entering an industrial process with optronic choices and signal processing completion, incorporating the energy constraints and limited electrical/optical or optical/electrical conversion. Optimized data transfers even in emergency situations together with a network without coordinator have already been implemented in a protocol suitable to wireless optic (OWMAC).

Because, as an indoor network solution in a home or an office environment, this communication medium has many attractive opportunities:

– Visible and infrared spectral regions provide a bandwidth that is twice the size of the radio field, more than 700,000 GHz actually unregulated and untaxed.

– Optical transmissions are limited inside a room. This feature allows us to physically and easily secure each communication. It is also possible to use the same optical wavelength in the neighboring room or apartment with the same level of security. This facilitates and makes the network architecture intuitive.

– Because of the spectral range used, there is no possibility of suffered or produced interference with radio systems in the vicinity. Moreover, it is also possible to convert a portion of this photon energy into electrical energy.

Eventually, we can imagine an even more prospective scenario with access and connectivity between rooms. Wireless optical communication is achieved between a station per room and one or more smart objects in the room whose characteristics and format are mobile phone and laptop.

This objective using wireless optical communication with a positive energy balance provides the features mentioned above, together with a universal remote control, a pointing and localization device to provide an appropriate interface in augmented reality or mastered immersion in virtual reality.

This global photonic approach coupled with the depletion of rare earths [BIH 10] and with the development of biophotonic solutions [ZYS 10] will be at the crossroads of the world of services to people, home automation, telecommunications, and computer science.

APPENDICES

Appendix 1

Geometrical Optics, Photometry and Energy Elements

A1.1. Geometrical optics elements

A1.1.1. *Refractive index*

The refractive index characterizes the ratio of the speed in vacuum (c_v) to the speed in another homogeneous environment (c_e). In the latter environment, the optical wave propagates at a speed lower than that in vacuum c_v. The ratio is expressed by the following equation:

$$n = \frac{c_v}{c_e}$$

Equation A1.1. *Refractive index*

Table A1.1 presents some refractive index examples.

Environment	Refractive index
Vacuum	1
Air	1.00029
Water	1.3333
Glass	1.5
Diamond	2.41

Table A1.1. *Examples of refractive indexes*

When the optical wave passes from one environment to another, not only its velocity (speed) but also its trajectory changes. In addition, this index varies with wavelength, which can cause optical aberrations (e.g. the phenomenon of light scattering in a prism).

A1.1.2. *Fermat's principle*

Fermat's principle is the basis for any geometrical optics that describes the refraction and reflection laws. Under this principle, the path followed by a light beam between two points A and B is such that the travel time between these two points is minimum. Then the light propagation is straight in a homogeneous medium with constant velocity (in classical Euclidean geometry). This feature also applies in the context of a path from B to A (principle of light reversibility).

As such, a set of optical waves form an optical beam whose trajectories fall into three categories:

– Parallel beam: the rays are parallel.

– Divergent beam: the rays originate from a single point.

– Convergent beam: the rays propagate toward the same point.

A1.1.3. *Snell–Descartes' laws*

When a ray passes from a homogeneous environment to another homogeneous environment with a different refractive index, part of this ray is reflected, and the other part is transmitted toward the second environment.

Snell–Descartes' laws describe the trajectory of these rays (case of the crossing from an n_1 index environment to an n_2 index environment with $n_1 < n_2$).

– Descartes' first law: the incident, reflected, and transmitted rays are contained in the same plane.

– Descartes' second law: the angles for the incident ray (i) and the reflected ray (r) are equal.

– Descartes' third law: regarding the refracted/transmitted ray: the rule is as follows: $n_1.\sin i_1 = n_2.\sin t$. If the n_1 index environment is air ($n_1 = 1$), then the refractive index n_2 of the second medium is the ratio of sinus of the angles i and t.

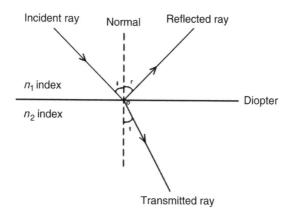

Figure A1.1. *Snell–Descartes' laws*

Note that an absorption phenomenon may occur in case the environment is not completely transparent.

For a mirror, there is no transmission, the ray is completely reflected. In the case of a glass diopter (lens, prism, etc.), depending on the surface treatment, about 4% of the wave is reflected and the rest is transmitted.

A1.1.4. *Definitions: sources, image, and focal point*

Source: the object from which the optical rays originate.

Spot object or spot source: a dimensionless source – or equivalent (a star, for example). Even if it bears a nonzero size, its distance allows it to be considered as a spot source.

Extended object or source: a nonzero size source. The rays are emitted from different points of the source (the Sun, for example).

Real object: a real object is a spot or an extended source from where divergent rays radiate.

Real image: a real image is the area of space where the light rays from one source converge after passing through an optical system. A real image can be projected on a screen.

Virtual object: an object or a virtual image is the area of space corresponding to the imaginary extension of optical rays. The virtual object is downstream from an optical device, while the virtual image is upstream. A virtual image cannot be projected on a screen.

Focal point: any light beam (from infinity) parallel to a converging lens' optical axis converges to a point called the focal point (F). Conversely, any image at the focal point for a diverging lens seems to come from infinity (inverse return of light). The focal point position is determined by the geometry of the lens and its refractive index.

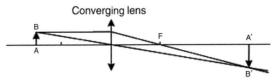

Figure A1.2. *Real object – real image*

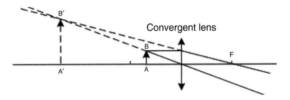

Figure A1.3. *Real object – virtual image*

A1.2. Photometry elements

Optical radiometry or photometry consists of measurement of the energy flux associated with optical electromagnetic radiations. The basic elements are flux, intensity, illumination, and luminance. An optical radiation source emits a luminous flux in potentially all directions of space. This flux has an intensity value for any given direction. A surface, located at a defined distance from the source, receives a given irradiance value.

Before going further, it may be wise to recall the definition of some physical quantities in the fields of visual and energetic photometry.

A1.2.1. *Steradian*

A steradian (sr) is a unit of solid angle which, its apex located at the center of a sphere, has on the surface of the sphere an area equivalent to that of a square whose side is equal to the radius of this sphere. It is a unit of the International System (IS). For a general sphere of radius r, any portion of its surface with area $A = r^2$ subtends 1 sr.

A1.2.2. *Solid angle*

The solid angle measures the apparent size under which an object appears to an observer. This is the ratio of the surface of the object projection on a sphere to its radius squared. For elementary surface dS, the elementary solid angle $d\Omega$ can be written as:

$$d\Omega = \frac{dS \, \cos\theta}{d^2}$$

Equation A1.2. *Solid angle*

where

– θ: angle between the normal to the elementary object and the direction of observation;

– dS: projection area;

– d: distance.

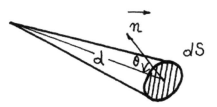

Figure A1.4. *Solid angle*

A 1 sr solid angle demarcates on a unit sphere, from the center of this sphere, a surface area equal to 1. For instance, the solid angle, defining a sphere or an observation on the entire space, is 4π sr and the solid angle of the Sun as seen from the Earth is 6×10^{-5} sr.

A1.2.3. *Light intensity*

The light intensity is the first physical quantity of visual photometry and is measured in candela (cd). The candela is the luminous intensity unit in the IS. One candela is equal to 1/683 W/sr of a monochromatic source at 540×10^{12} Hz (555 nm). This wavelength corresponds to the yellow–green color.

A1.2.4. *Luminous flux*

The luminous flux is the second physical quantity of visual photometry; and is expressed in lumens (lm). One lumen corresponds to a flux emitted by a light source with a 1 cd intensity in a 1 sr solid angle. 1 lm = 1 cd/1 sr.

A1.2.5. *Illumination*

Illumination is the third physical quantity of visual photometry; the lux (lx) is the unit of illumination. One lux is the illumination of an area located 1 m away from a 1 cd light source. $1 \text{ lx} = 1 \text{ lm/m}^2$.

Table A1.2 presents some usual illumination values.

Night sky	0.0003
Full Moon	0.2
75 W electric lamp at 2 m	40
Public lighting	50
Cloudy weather	15,000
Sunlight in summer at noon	100,000

Table A1.2. *Usual illumination values*

A1.2.6. *Luminance*

The luminance is the fourth physical quantity of visual photometry and is the only quantity noticeable by a human eye. The luminance (cd/m^2) is the result of the ratio of the luminous intensity of a remote source (e.g. 1 cd) to the value of the apparent area in a given direction (e.g. 1 m^2).

Table A1.3 presents some usual luminance values.

Eye perception threshold	0.000001
Full Moon	2,000
White paper in sunlight in summer at noon	30,000
Projection lamp filament	20,000,000
Threshold for potential eye injury	250,000,000
Sun through the atmosphere	1,600,000,000
Flashlamp	10,000,000,000

Table A1.3. *Usual luminance values*

A1.2.7. *Conversion of visual photometry into energetic photometry*

As a principle, photometric values are converted from energy units to visual units by weighting the values with the spectral sensitivity curve of the eye.

For a monochromatic beam (narrow spectral band), the luminous flux $F_v(\lambda)$ at wavelength λ is expressed by the following equation:

$$F_v(\lambda) = K_{e\lambda} F_e(\lambda) = K_m V(\lambda) F_e(\lambda)$$

Equation A1.3. *Monochromatic luminous flux*

where:

– $F_v(\lambda)$: luminous flux in lumen;

– $F_e(\lambda)$: energetic flux in watt;

– $K_{e\lambda}$: luminous spectral efficiency;

– $V(\lambda)$: relative spectral sensitivity of the eye (CIE standard) for wavelength λ.

By convention, the curve $V(\lambda)$ is defined from 380 to 780 nm. The maximum sensitivity is at $\lambda = 555$ nm (maximum of the CIE curve $V(\lambda)$) and the K_e value equals 683 lm/W.

For example, if we are looking for the eye sensitivity value in energy photometry, we follow the following approach. To simplify the calculation, we consider the luminance of a source at 555 nm with an apparent 1 m² surface area under 1 sr angle.

The luminous flux becomes 0.000001 lm with $V(\lambda) = 1$ and $K_m = 683$ lm/W.

$$F_e(555) = \frac{F_v(555)}{K_m V(555)} = \frac{0.000\,001}{683} = 1.5 \times 10^{-9}$$

The eye sensitivity value in energy photometry (at 555 nm) is 1.5 nW, or about −60 dBm.

For a polychromatic beam (wide spectral band), this relationship is:

$$dF_v = K_{e\lambda}dF_e = K_m V(\lambda)dF_e = K_m V(\lambda)\left[\frac{dF_e}{d\lambda}\right]d\lambda = \frac{dF_v}{d\lambda}d\lambda$$

The global visual flux becomes:

$$F_v = \int_0^\infty dF_v = K_m \int_{380}^{780} V(\lambda)\frac{dF_e}{d\lambda}d\lambda$$

Equation A1.4. *Polychromatic luminous flux*

And the global energy flow is:

$$F_e = \int_0^\infty \frac{dF_e}{d\lambda}d\lambda$$

Equation A1.5. *Global energy flux*

A1.2.8. Bouguer's relation

The Bouguer's relation expresses the variation in illumination depending on light intensity, view angle, and distance:

$$E = \frac{I\cos\theta}{d^2}$$

Equation A1.6. *Bouguer's relation*

This relation shows that the illumination for a *dS* surface, located at distance *d* from a spot-light source *I*, varies inversely as the squared distance. So, a surface located twice as far from the source has four times greater the area but four times lower the illumination.

A1.2.9. *Energy flux or radiated power P*

Any optical radiation transports energy at a given speed in the propagation medium. This energy is expressed in joule (J). When it is defined per time unit, it is called energy flux, emitted flux or received power at the considered time. This quantity is the temporal flux of energy and is expressed in watt (W).

In the field of optical wireless devices, this energy flux is characterized by several parameters: average power, peak or maximum power, and operation mode (continuous or pulsed at a determined frequency). To transmit information, it is possible to vary the radiation power or its amplitude. This is then in any case a pulsed mode optical flux, with specific power and energy characteristics.

$$P_0 = P_c \frac{t}{T}$$

Equation A1.7. *Average power in pulsed mode*

where:

– P_0: average power (W);

– P_c: peak power (W);

– t: pulse duration (s) ;

– T: signal period (s).

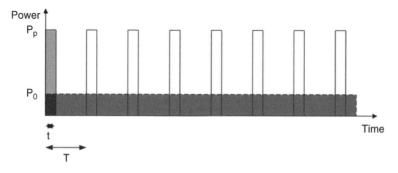

Figure A1.5. *Pulsed mode*

The relationship between the power value in W and dBm is given by:

$$P_{dBm} = 10 \log\left(\frac{P_w}{0.001}\right)$$

Equation A1.8. *Power (dBm)*

A1.2.10. *Source intensity I*

The intensity I for a source in a given direction is the emitted flux from the source per solid angle unit in a considered direction. The intensity is expressed in watt per steradian (W/sr).

If dF_s is the flux emitted by the source in the elementary solid angle $d\Omega_s$, the intensity of the source in this direction is:

$$I = \frac{dF_s}{d\Omega_s}$$

Equation A1.9. *Intensity of a source*

A1.2.11. *Luminance of a source L*

To characterize the radiation from a source in more detail than the simple knowledge of its flux or intensity, the source can be decomposed into a set of small dA_s area elements, independent of each other. The elementary intensity dI corresponding to dA_s is defined by the following expression:

$$dI = \frac{d^2 F_s}{d\Omega_s}$$

where $d^2 F_s$ represents the flux radiated by the small dA_s area element in the solid angle $d\Omega_s$.

The energy luminance for a dA_s area source is the source intensity per apparent area unit in that direction. Mathematically, this gives:

$$L = \frac{dI}{dA_s \cos\theta_s}$$

where θ_s is the angle between the local normal to the source and the emission direction, hence:

$$L = \frac{d^2 F_s}{dA_s \cos \theta_s d\Omega_s}$$

Equation A1.10. *Energy luminance*

The luminance energy unit, for which the symbol is L, is the watt per square meter per steradian (W/m^2/sr).

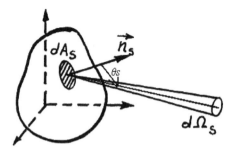

Figure A1.6. *Luminance of the source*

A1.2.12. *Illumination of a receiving surface E*

The surface's irradiance E at a receiver's point is the incident energy flux dF_r received on a surface element dA_r per area unit and limited by the projection $dA_r.\cos\theta_r$, with θ_r which is the angle between the local normal to the reception surface and the emission direction:

$$E = \frac{dF_r}{dA_r \cos \theta_r}$$

Equation A1.11. *Illumination of receiving surface E*

The irradiance unit is the watt per square meter, W/m^2.

A1.2.13. *Geometrical extent G*

In the case of a light beam, the expression for the source luminance can be rewritten as:

$$d^2 F_s = L \, dA_s \cos \theta_s d\Omega_s = L \, d^2 G$$

The quantity $d^2 G = dA_s \cos \theta_s d\Omega_s$ characterized by the source emissive surface and the emission solid angle is called the "geometrical extent" of the light beam from the emitter.

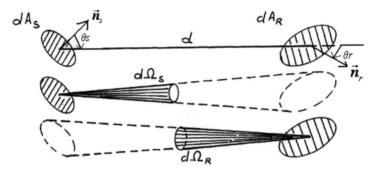

Figure A1.7. *Geometrical extent of a light beam*

In Figure A1.7, the emission solid angle for this beam is defined by the silhouette of a receiver of elementary area dA_r located at distance d from the source:

$$d\Omega_s = \frac{dA_r \cos \theta_r}{d^2}$$

We note that the geometrical extent of a beam defined by two diaphragms of areas dA_s and dA_r located at the mutual distance d can be expressed in two other ways:

$$d^2 G = dA_s \cos \theta_s d\Omega_s = dA_s \cos \theta_s \frac{dA_r \cos \theta_r}{d^2} = dA_r \cos \theta_r d\Omega_r$$

Equation A1.12. *Light beam*

where $d\Omega_r$ is the solid angle for the source element dA_s as seen from the second diaphragm dA_r.

In the case of a light beam, the geometrical extent G is calculated by summing the elementary geometrical extent contributions of all constituted beams.

A1.3. Equivalence between visual and energetic photometry

Energetic photometry		Visual photometry	
IS radiometry units		IS photometry unit	
Magnitude	Unit	Correspondence	Unit
Power (P)	Watt (W)	Luminous flux (F)	Lumen (lm)
Energy	Joule (J)	Light quantity	Lumen.second (lm.s)
Energetic intensity (I)	Watt per steradian (W/sr)	Light intensity(I)	Candela (cd)
Energetic luminance (L)	Watt per steradian and per square meter (W/sr/m^2)	Luminance (brightness) (L)	Candela per square meter (cd/m^2)
Energetic illumination (E)	Watt per square meter (W/m^2)	Illumination (H)	Lux (lx)

Table A1.4. *Equivalence table between radiometry and photometry*

Appendix 2

The Decibel Unit (dB)

The decibel (dB) is one tenth of the Bel unit, defined in honor of Alexander Graham Bell. It is a logarithmic unit of measurement of the ratio of two powers, voltages, currents, pressures, etc. It is particularly used in the fields of acoustics, physics and electronics, etc.

This unit of measurement (dB) is defined by the logarithmic ratio:

– of potential (V) or current (I):

$$-20\log (V_{output} / V_{input})$$

$$20\log (I_{output} / I_{input})$$

– of power (P):

$$-10\log (P_{output} / P_{input})$$

If the result is positive, this is a case of amplification. If the result is negative, this is a case of attenuation.

NOTE:– The term "log" refers here to the base 10 logarithm.

There are different dB variants: dBm, dBW (dB watt), dBi, dBd, dBc, dBμV (dB microvolt), dBμVm[1] (dB microvolt/m):

– dB: $-10\log_{10} (P_{output} / P_{input})$.

– dBm: $-10\log_{10} (P_{output} / 1\text{mW})$.

– dBW: $-10\log_{10}(P_{output}/1\,\text{W})$.

– dBi: the antenna's gain versus the gain for an isotropic antenna. An isotropic antenna is an antenna that emits the same amount of energy in all directions. Such an antenna does not exist in reality.

– dBd: the antenna's gain versus the gain for a half-wave dipole antenna. This antenna has a physical reality.

Note that catalogs often do not specify whether we are dealing with dBi or dBd. Yet, the difference is important. A 10 dBd antenna gain is a 12.15 dBi antenna gain. We see that the temptation is great to display dBi rather than dBd without explaining it.

– dBc: $-10\log_{10}(P_{output}/P_c)$. Here, an output power (P_{output}) is compared to that of a carrier (P_c) where c is the carrier.

EXAMPLE: Amplitude difference between the fundamental signal and the various harmonics. Reduced to 1 Hz, we refer to dBc/Hz (dB relative to the carrier per Hz of bandwidth).

– dBμV: $-20\log_{10}(voltage/1\,\mu\text{W})$. The reference voltage is equal to 1 μV.

– dBμV/m: $-10\log_{10}(field/1\,\eta\text{Vm}^{-1})$. The reference field is 1 μV/m^1.

The various operations on decibels are summarized below:

$$\log(ab) = \log a + \log b$$
$$\log(a/b) = \log a - \log b$$
$$\log(a^b) = b\log a$$

If

$$A(dB) = 10\log_{10}\left(\frac{P_s}{P_e}\right) = 10\log_{10}(R)$$

then

$$R = 10^{\frac{A}{10}}$$

NOTE: The advantage of using the dB scale is that gains (amplifications) or attenuations are added (instead of multiplied).

The following table shows some correspondence between numerical values and their dB values.

R	dB
100,000 (10^5)	50
10,000 (10^4)	40
1,000 (10^3)	30
100 (10^2)	20
10 (10^1)	10
1 (10^0)	0
0.1 (10^{-1})	−10
0.01 (10^{-2})	−20
0.001 (10^{-3})	−30
0.0001 (10^{-4})	−40
0.00001 (10^{-5})	−50
0.000001 (10^{-6})	−60
0.0000001 (10^{-7})	−70
0.00000001 (10^{-8})	−80
0.000000001 (10^{-9})	−90

The dB values for some particular numerical values (10, 2, 0.1, and 0.5) are listed below:

– $10*\log(10) = 10$;

– $10*\log(2) = 3$;

– $10*\log(1/10) = -10$;

– $10*\log(1/2) = -3$.

As a result, the rules to remember are the following:

– Whenever we multiply by 10 in the current scale, we add 10 (dB) in the decibel scale.

– Whenever we multiply by 2 in the current scale, we add 3 (dB) in the decibel scale.

– Whenever we divide by 10 in the current scale, we subtract 10 (dB) in the decibel scale.

– Whenever we divide by 2 in the current scale, we subtract 3 (dB) in the decibel scale.

EXAMPLE A2.1:–

- 10 mW + 3 dB = 20 mW;
- 100 mW – 3 dB = 50 mW;
- 10 mW + 10 dB = 100 mW;
- 100 mW – 10 dB = 10 mW.

Bibliography

[ACH 04] ACHOUR M., "Free-space optical communication by retro-modulation: concept, technologies, and challenges", *Proceedings of SPIE*, vol. 5614, 2004.

[ALH 06] ALHAMMADI K., Applying wide field of view retro-reflector technology to free space optical robotic communications, PhD Thesis, North Carolina State University, 2006.

[ALQ 03a] ALQUDAH Y., Space diversity techniques in indoor broadband optical wireless communications, PhD Thesis, Pennsylvania State University, 2003.

[ALQ 03b] ALQUDAH Y.A., KAVEHRAD M., "MIMO characterization of indoor wireless optical link using a diffuse-transmission configuration", *IEEE Transactions on Communications*, vol. 51, no. 9, pp. 1554–1560, September 2003.

[ALQ 04] ALQUDAH Y., KAVEHRAD M., JIVKOVA S., "Optical wireless multi-spot diffusing: a MIMO configuration", *Proceedings of ICC 2004*, Paris, 2004.

[AUD 96] AUDEH M.D., KAHN J.M., BARRY J.R., "Performance of pulse-position modulation on measured non-directed indoor infrared channels", *IEEE Transactions on Communications*, vol. 44, pp. 654–659, June 1996.

[BAH 74] BAHL L.R., COCKE J., JELINEK F., RAVIV J., "Optimal decoding of linear codes for minimizing symbol error rate", *IEEE Transactions on Information Theory*, vol. 20, pp. 284–287, March 1974.

[BAR 93] BARRY J.R., KAHN J.M., KRAUSE W.J., LEE E.A., MESSERSCHMITT D.G., "Simulation of multipath impulse response for indoor wireless optical channels", *IEEE Journal on Selected Areas in Communications*, vol. 11, pp. 367–379, April 1993.

[BAR 94] BARRY J.R., *Wireless Infrared Communications*, Kluwer Academic Publishers, Boston, 1994.

[BAT 92] BATAILLE P., Analysis of the behavior of a telecommunications system operated optical 0.83 microns in the lower atmosphere, PhD thesis, University of Rennes, France, 1992.

[BEL 10] BELMONTE A., KAHN J.M., "Field conjugation adaptive arrays in atmospheric coherent optical link", *IEEE Globecom Miami Workshop*, OWC2#5, December 2010.

[BEN 92] BENNET C.H., BESSETTE F., BRASSARD G., SALVAIL L., SMOLIN J., "Experimental quantum cryptography", *Journal of Cryptography*, vol. 5, no. 3, pp. 1992.

[BER 68] BERLEKAMP E.R., *Algebraic Coding Theory*, chapter 7, McGraw-Hill, New York, 1968.

[BER 93] BERROU C., GLAVIEUX A., THITIMAJSHIMA P., "Near Shannon limit error-correcting coding and decoding: turbo-codes", *IEEE International Conference on Communications*, ICC 1993, vol. 2, pp. 1064–1070, May 1993.

[BER 08] BERTRAND M., BOUCHET O., EL TABACH M., BESNARD P., LEROUX D., "Qualité Optique sans Fil Indoor", free software, December 2008, available at http://departements. telecom-bretagne.eu/sc/recherche/techimages.

[BES 10] BESNARD P., FAVENNEC P.N., *Les applications du laser – 50 ans après son invention*, Collection Télécom, Hermès-Lavoisier, Paris, 2010.

[BEV 02] BEVERATOS A., BROURI R., GACOIN T., VILLING A., POIZAT J.P., GRANGIER P., "Single photon quantum cryptography", *Physical Review Letters*, vol. 89, no. 18, 2002.

[BIG 10] BIGO S., "Les très hauts débits", *Workshop Lasers et communication*, Institut Télécom, Paris, December 2010.

[BIH 10] BIHOUIX P., DE GUILLEBON B., *Quel futur pour les métaux*, EDP sciences, Les Ulis, 2010.

[BOD 77] BODEUX A., "La fréquence des brouillards en Belgique", Institut royal météorologique de Belgique, Brussels, 1977.

[BOS 60] BOSE R.C., RAY-CHAUDHURY D.K., "On a class of error correcting binary group fields", *Information and Control*, vol. 3, pp. 68–79, March 1960.

[BOU 04] BOUCHET O., SIZUN H., BOISROBERT C., DE FORNEL F., FAVENNEC P.N., *Optique sans fil, propagation et communication*, CTST, Hermès-Lavoisier, Paris, 2004.

[BOU 07] BOUCHET O., ROUET C., "Indoor optical wireless communication: interference experimentation and evaluation", *SPIE Optics and Photonics Conference*, vol. 6304, San Diego, USA, 2007.

[BOU 08] BOUCHET O., BERTRAND M., BESNARD P., "Personal optical wireless communications: LOS/WLOS/DIF propagation model and QOFI", *IEEE International Symposium on Communication Systems, Networks and Digital Signal Processing (IEEE CSNDSP)*, July 2008.

[BOU 09] BOUCHET O., LAUNAY G., BESNARD P., "Indoor optical wireless communication: a giga Ethernet network at 60 dB link margin", *SPIE Optics and Photonics Conference*, vol. 7464, August 2009.

[BOU 09a] BOUCHET O., LAUNAY G., Notes techniques – Système de transmission optique : Rapport de mesures, Projet Techim@ges, Pôle de compétitivité Images et réseaux, March 2009.

[BRU 92] BRUHAT G., Cours de Physique Générale, 6th ed., Masson, Paris, 1992.

[CAR 98] CARBONNEAU T.H., WISELEY D.R., "Opportunities and challenges for optical wireless; the competitive advantage of free space telecommunications links in today's crowded market place", SPIE Conference on Optical Wireless Communications, vol. 3232, Boston, USA, 1998.

[CAR 00] CARRUTHERS J.B., KAHN J.M., "Angle diversity for nondirected wireless infrared communication", IEEE Transactions on Communications, vol. 48, no. 6, pp. 960–969, 2000.

[CAR 02a] CARRUTHERS J.B., "Wireless infrared communications", Wiley Encyclopedia of Telecommunication, Wiley, New York, USA, 2002.

[CAR 02b] CARRUTHERS J.B., KANNAN P., "Iterative site-based modeling for wireless infrared channels", IEEE Transactions on Antennas and Propagation, vol. 50, pp. 759–765, 2002.

[CHA 05] CHABANE M., AL NABOULSI M., SIZUN H., BOUCHET O., "A new quality of service FSO software", European Conference on Propagation and Systems (ECPS '05), Brest, France, 2005.

[CIM 85] CIMINI L.J., "Analysis and simulations of a digital mobile channel using orthogonal frequency division multiplexing", IEEE Transactions on Communications, vol. 33, no. 7, pp. 665–675, July 1985.

[CLA 81] CLAY M.R., LENHAM A.P., "Transmission of electromagnetic radiation in fogs in the 0.53–10.1 µm wavelength range", Applied Optics, vol. 20, no. 22, pp. 3831–3833, 1981.

[COJ 99] COJAN Y., FONTANELLA J.C., Propagation du rayonnement dans l'atmosphère, Technique de l'ingénieur, Traité électronique, 1999.

[COZ 83] COZANNET A., FLEURET J., MAITRE H., ROUSSEAU M., Optique et télécommunications, Collection Technique et Scientifique, Eyrolles, Paris, 1983.

[DAV 09] DAVIS C.C., ESLAMI M., AGRAWAL N., "Channel modelling for FSO communications and sensor networking inside structures", SPIE Optics and Photonics, vol. 7464, August 2009.

[DEF 01] DE FORNEL F., FAVENNEC P.N., RAHMANI A., SALOMON L., BERGUIGA L., Source à peu de photons commandables, French patent 00 03-96, 2000; Single photon source and sources with few photons, US patent WO 01/69841 A1, 2001.

[DEF 02] DE FORNEL F., "Matériaux artificiels pour les communications sécurisées", in C. KINTZIG et al. (ed.), Les objets communicants, CTST, Hermès-Lavoisier, Paris, 2002.

[DEF 07] DE FORNEL F., FAVENNEC P.N., "Mesures en électromagnétisme", revue I2M, vol. 7, Hermès-Lavoisier, Paris, 2007.

[DEF 10] DE FORNEL F., FAVENNEC P.N., *Measurements Using Optic and RF Waves*, ISTE, Wiley, London, 2010.

[DEI 69] DEIRMENDJIAN D., *Electromagnetic Scattering on Spherical Polydispersions*, Elsevier, New York, 1969.

[DJA 00] DJAHANI P., KAHN J.M., "Analysis of infrared wireless links employing multibeam transmitters and imaging diversity receivers", *IEEE Transactions on Communications*, vol. 48, pp. 2077–2088, December 2000.

[DLI 05] D-DLINK 802.11(g) PCMIA card, 2005.

[ENE 11]   www.energies-nouvelles.net/outil-calcul-rendement-panneau-photovoltaique-4.html, 2011.

[ELT 09] EL TABACH M., A contribution to digital communications studies for optical wireless networks, PhD Thesis, Orange Labs, TELECOM Bretagne, Rennes, 2009.

[EN 07] EUROPEAN STANDARD EN 60825-1, Safety of laser products – part 1: equipment classification and requirements, 2nd ed., 2007.

[EN 03] EUROPEAN STANDARD EN 12464-1, Lighting of indoor work places, 2003.

[FUN 02] FUNK J., KNUTSON C.D., "Evaluating the capacity of RLL encoding for high bandwidth infrared channels", *Proceedings of the Third International Symposium on Communication Systems, Networks and Digital Signal Processing* (CSNDSP '02), Stafford, UK, 15–17 July 2002.

[GAL 63] GALLAGER R.G., *Low-Density Parity-Check Code*, University of Cambridge, Cambridge, UK, July 1963.

[GAR 04] GARRIDO J.M., GARCIA-ZAMBRANA A., PUERTA-NOTARIO A., "Performance evaluation of rate adaptive transmission techniques for optical wireless communication", *VTC 2004-Spring, 59th Vehicular Technology Conference*, Milan, Italy, 2004.

[GEB 04] GEBBART M., LEITGEB E., AL NABOULSI M., SIZUN H., DE FORNEL F., "Measurements of light attenuation at different wavelengths in dense fog conditions for FSO applications", *STSM-7, COST270*, Brussels, Belgium, 2004.

[GFE 79] GFELLER F.R., BAPST U., "Wireless in-house data communications via diffuse infra red radiations", *IEEE Proceedings*, vol. 67, no. 11, pp. 1474–1486, 1979.

[GIG 02] GIGGENBACH D., PURVINSKIS R., WERNER M., HOLZBOCK M., "Optical inter-plaform links for high altitude platforms", *AIAA, Proceeding of the 20th International Communications Satellite Systems Conference (ICSSC)*, Montréal, Canada, May 2002.

[GLA 96] GLAVIEUX A., JOINDOT M., *Communications Numériques, Introduction*, Masson, Paris, 1996.

[GON 05] GONZALEZ O., PEREZ-JIMENEZ R., RODRIIGUEZ S., RABADAN J., AYALA A., "OFDM over indoor wireless optical channel", *IEEE Proceedings*, pp. 199–204, 5 July 2005.

[GOR 84] GORAL C.M., TORRANCE K.E., GREENBERG D.P., BATTAILE B., "Modelling the interaction of light between diffuse surfaces", *Computer Graphics*, vol. 18, no. 3, July 1984.

[GRA 07] GRABNER M., KVICERA V., "On the relation between atmospheric visibility and wave attenuation", *16th IST Mobile and Wireless Communication Summit*, Budapest, Hungary, 2007.

[HAE 03] HAESE-COAT V., KPALMA K., *Traitement numérique du signal*, Ellipses, Paris, 2003.

[HAP 11] HIGH ALTITUDE PLATFORMS FOR COMMUNICATIONS AND OTHERS SERVICES, COST 297, 2011, available at www.hapcos.org/DOCS/wg2/wg2_home.php.

[HAR 94] HARDING G.F.A., JEAVONS P., *Photosensitive Epilepsy*, MacKeith Press, London, UK, 1994.

[HAS 94] HASHEMI H., YUN G., KAVEHRAD M., BEHBAHANI F., GALKO P.A., "Indoor propagation measurements at infrared frequencies for wireless local area networks applications", *IEEE Transactions on Vehicular Technology*, vol. 43, no. 3, pp. 562–576, August 1994.

[HAU 05] HAUTIÈRE N., AUBERT D., JOURLIN M., "Détection des conditions de visibilité et estimation de la distance de visibilité par vision embarquée", *MajecSTIC*, Rennes, France, 2005.

[HAU 06] HAUTIÈRE N., AUBERT D., JOURLIN M., "Mesure du contraste local dans les images, application à la mesure de la distance de visibilité par caméra embarquée", *Traitement du signal*, vol. 23, no. 2, pp. 145–158, 2006.

[HAY 10] HAYATA N., MIYAMOTO S., SAMPEI S., A proposal of intensity distribution-based multiplexing scheme for downlink LOS indoor optical wireless communication system, IEICE Technical Report, RCS2010-41 (2010-6), 2010.

[HEN 05] HENNIGER H., GOGGENBACH D., RAPP C., "Evaluation of optical up and downlinks from high altitude platforms using IM/DD", *Proceedings of the SPIE-Conference on Free Space Laser Communication Technology*, San Jose, USA, January 2005.

[HEN 10] HENNIGER H., WILFERT O., "An introduction to free space optical communications", *Radioengineering*, vol. 19, no. 2, June 2010.

[HIR 01] HIRT W., HASSNER M., HEISE N., "IrDA-VFIr (16 mb/s): modulation code and system design", *IEEE Personal Communication*, vol. 8, February 2001.

[HOR 04] HORWATH J., KNAPEK M., PERLOT N., GIGGENBACH D., "Optical communication from HAPs – overview of the stratospheric optical payload experiment (STROPEX)", *AIAA, Proceedings of the 22th ICSSC*, Monterrey, Mexico, May 2004.

[IAR 11] http://www.iarc.fr/en/media-centre/pr/2011/pdfs/pr208_E.pdf.

[ICS 11] INFRARED COMMUNICATION SYSTEMS ASSOCIATION, 2011.

[IEE 11a] IEEE 802.15 Wireless Personal Area Networks (WPAN), TG7, Visible Light Communication, available at www.ieee802.org/15/pub/TG7.html, 2011.

[IEE 11b] IEEE802.11 Wireless Local Area Networks (WLAN), 2011, available at http://ieee802.org/11/.

[IRD 11] INFRARED DATA ASSOCIATION, 2011, available at www.irda.org/.

[ITU 03] ITU -R P.1622, Prediction methods required for the design of earth-space systems operating between 20 THz and 375 THz, 2003.

[ITU 04] ITU -R P.837-4, Characteristics of precipitation for propagation modeling, 2004.

[ITU 05a] ITU-R 237/5, Preliminary Draft Revision of Question, Fixed service applications using frequency bands above 3 000 GHz, Document 5C/TEMP/34-E, 12 February 2005.

[ITU 05b] ITU -R P.1621-1, Propagation data required for the design of earth-space systems operating between 20 THz and 375 THz, 2005.

[ITU 07a] ITU -R P.1817, Propagation data required for the design of terrestrial free-space optical links, 2007.

[ITU 07b] ITU -R P.1814, Prediction methods required for the design of terrestrial free-space optical links, 2007.

[ITU 08] PDNR ITU-R F.2106, Fixed service applications using free-space optical links, Document 5C/129, 17 November 2008.

[ITU 11] ITU, Internationale Telecommunications Union, 2011, available at www.itu.int/fr/pages/default.aspx.

[JIV 01] JIVKOVA S., KAVEHRAD M., "Holographic optical receiver front end for wireless infrared indoor communications", *Applied Optics*, vol. 40, pp. 2828–2835, 2001.

[JIV 04] JIVKOVA B., HRISTOV B.A., KAVEHRAD M., "Power-efficient multispot-diffuse multiple-input–multiple-output approach to broad-band optical wireless communications", *IEEE Transactions on Vehicular Technology*, vol. 53, no. 3, May 2004.

[JOI 96] JOINDOT I., JOINDOT M., *Les télécommunications par fibres optiques*, Eyrolles, Paris, 1996.

[KAH 95] KAHN J.M., KRAUSE W., CARRUTHERS J., "Experimental characterization of non-directed indoor infrared channels", *IEEE Transactions on Communications*, vol. 43, nos. 2–4, February–April 1995.

[KAH 97] KAHN J.M., BARRY J.R., "Wireless infrared communications", *IEEE Proceedings*, vol. 85, no. 2, pp. 265–298, 1997.

[KAH 98] KAHN J.M., YOU R., DJAHANI P., WEISBIN A.G., TEIK B.K., TANG A., "Imaging diversity receivers for high-speed infrared wireless communication", *IEEE Communications Magazine*, December 1998.

[KDD 11] USB 2.0 Wireless Infrared Communication Systems manufactured by KDDI operator, 2011.

[KIM 01] KIM I.I., MCARTHUR B., KOREVAAR E.J., "Comparison of laser beam propagation at 785 nm and 1550 nm in fog and haze for optical wireless communications", *Proceedings of SPIE*, vol. 4214, pp. 26–37, 2001.

[KIN 02] KINTZIG C., POULAIN G., PRIVAT G., FAVENNEC P.N., *Objets Communicants*, CTST, Hermès-Lavoisier, Paris, 2002.

[KRU 62] KRUSE P.W., MCGLAUCHLIN L.D., MCQUISTAN R.B., *Elements of Infrared Technology: Generation, Transmission and Detection*, Wiley & Sons, New York, 1962.

[KÜL 98] KÜLLER R., LAIKE T., "The impact of flicker from fluorescent lighting on well-being, performance and physiological arousal", *Ergonomics*, vol. 41, no. 4, pp. 433–447, 1998.

[KUR 02] KURTSIEFER C., ZARDA P., HALDER M., WEINFURETER H., GORMAN P.M., TAPSTER P.R., RARITY R., "A step towards global key distribution", *Nature*, vol. 41, no. 9, p. 450, 2002.

[KWO 92] KWONG P.W., BRUNO R.C., "Communication system, evolutionary scenarios for martian SEI support; final report prepared for NASA Lewis Research Center", *Stanford Telecom*, 12 February 1992.

[LAL 07] LALAM M., LEROUX D., Rapport d'avancement modulations et codages, Sous-Projet 4 Techim@ges, Internal report, ENST Bretagne, 2007.

[LAS 01] LAST M., PATEL S., FISHER B., EZEKWE C., HOLLAR S., LEIBOWITZ B., PISTER K.S.J., "Video semaphore decoding for free-space optical communication", *Proceedings of SPIE*, vol. 4303, 2001.

[LEE 95] LEE D.C., KAHN J.M., AUDEH M.D., "Performance of pulse-position modulation with trellis-coded modulation on non-directed indoor infrared channels", *Proceedings of IEEE GLOBECOM '95*, vol. 3, pp. 1830–1834, Singapore, November 1995.

[LEE 97] LEE D.C., KAHN J.M., "Trellis-coded pulse-position modulation for indoor wireless infrared communications", *IEEE Transactions on Communications*, vol. 45, no. 9, pp. 1080–1087, 1997.

[LEF 95] LE FLOCH B., ALARD M., BERROU C., "Coded orthogonal frequency division multiplex", *Proceedings of the IEEE*, vol. 83, pp. 982–996, June 1995.

[LEI 03] LEITGEB E., GEBHART M., FASSER P., BREGENZER J., TANCZOS J., "Impact of atmospheric effects in free space optics transmission systems", *Proceedings of SPIE*, vol. 4976, p. 86, 2003.

[LEO 91] LASER ELECTRO-OPTICS TECHNOLOGY (LEOT) curriculum materials from the Center for Occupational Research and Development (CORD), Waco, Texas, 1991.

[LOM 98] LOMBAY C.R., VALADAS R.T., DE OLIVEIRA DUARTE A.M., "Experimental characterisation and modelling of the reflection of infrared signals on indoor surfaces", *IEEE Proceedings Optoelectronics*, vol. 145, no. 3, pp. 191–197, 1998.

[LOH 11] LIAISON OPTIQUE HAUT DÉBIT MOBILE, 2011, available at www.pole-scs.org/scs_project20155.fr.htm.

[LOP 97] LOPEZ-HERNANDEZ F.J., BETANCOR M.J., "DUSTIN: algorithm for the calculation of impulse response on IR wireless indoor channels", *Electronics Letters*, vol. 33, no. 21, pp. 1804–1806, 1997.

[MAM 98] MAMORU F., ARAKI T., YAMAKAWA S., HISADA Y., KONDO F., AKIBA T., "Studies on small-size 2-axis mechanical devices for FPM and PAM using piezelectronic actuators", *Proceedings of SPIE, Free-Space Laser Communication Technologies X*, vol. 3266, 1998.

[MAR 96] MARSH G.W., KAHN J.M., "Performance evaluation of experimental 50-Mbit/s diffuse infrared wireless link using on-off keying with decision-feedback equalization", *IEEE Transactions on Communications*, vol. 44, no. 11, pp. 1496–1504, 1996.

[MAX 54] MAXWELL J.C., "Solution of problems", *Cambridge and Dublin Mathematics Journal*, vol. 8, pp. 188–195, 1854.

[MAY 05] MAYNARD J.A., BEGLEY D., "Airborne laser communication: past, present and future", *Proceeding of SPIE, Free-Space Laser Communications V*, vol. 5892, 2005.

[MIC 85] MICHELSON A.M., LEVESQUE A.H., *Error-Control Techniques for Digital Communication*, Wiley-Interscience, New York, USA, 1985.

[MIE 08] MIE G., "Beiträge zur Optik trüber Medien, speziell kolloidaler Metallösungen", *Annales de Physique*, 25, pp. 377–445, 1908.

[MIH 07] MIHAESCU A., BESNARD P., FÉRON P., BOUCHET O., TRAYNOR N., MONTEVILLE A., "Realizing optical amplifiers with micro-sphere: a tiny 15 dB Gain, 2 dB noise factor amplifier", *CLEO*, Munich, Germany, 2007.

[MOR 96] MOREIRA A., VALADAS R., DUARTE A., "Performance of infrared transmission systems under ambient light interference", *IEEE Proceedings Optoelectronics*, vol. 143, no. 6, pp. 339–346, December 1996.

[MRV 11] Supplier of optical communications equipment and services, 2011, available at www.mrv.com.

[NAB 04] AL NABOULSI M., SIZUN H., DE FORNEL F., "Fog attenuation prediction for optical and infrared waves", *Optical Engineering*, vol. 43, no. 2, pp. 319–329, 2004.

[NAB 05] AL NABOULSI M., Contribution à l'étude des liaisons optiques atmosphériques: propagation, disponibilité et fiabilité, Thesis, University of Bourgogne, Dijon, 2005.

[NET 99] NETTLE P., "Radiosity in English", 20 May 1999, available at www.paulnettle.com.

[NEW 18] NEWTON I., *Opticks of Light*, Innys ed., Royal Society, London, 1718.

[NIC 77] NICODEMUS F.E., RICHMOND J.C., HSIA J.J., Geometrical considerations and nomenclature for reflectance, Technical Report of NBS Monograph, vol. 160, National Bureau of Standards, US Department of Commerce, Washington, 1977.

[OBR 00] O'BRIEN D.C., FAULKNER G.E., KALOK J., ZYAMBO E.B., EDWARDS D.J., WHITEHEAD M., STRAVINAOU P., PARRY G., BELLON J., SIBLEY M.J., LALIHHAMBIKA V.A., JOYNER V.M., SAMDUSIN R.J., ATKINSON R., HOLBURN D.M., MEARS R., "High speed integrated optical wireless transceivers for in-building optical LANs", *Optical Wireless Communications III, SPIE*, Boston, USA, 2000.

[OBR 03] O'BRIEN D.C., FAULKNER G.E., JIM K., ZYAMBO E.B., EDWARDS D.J., WHITEHEAD M., STRAVINAOU P., PARRY G., BELLON J., SIBLEY M.J., LALIHHAMBIKA V.A., JOYNER V.M., SAMDUSIN R.J., HOLBURN D.M., MEARS R.J., "High-speed integrated transceivers for optical wireless", *IEEE Communications Magazine*, pp. 58–62, March 2003.

[OBR 04] O'BRIEN D., KATZ M., "Short-range optical wireless communications", *WWRF*, 2004.

[OBR 05] O'BRIEN D.C., "Improving coverage and data rate in optical wireless systems", *Free-Space Laser Communication V*, San Diego, USA, 2005.

[OME 08a] Omega ICT Deliverable D4.1, State-of-the-art HWO, October 2008, available at www.ict-omega.eu/publications/deliverables.html.

[OME 08b] Omega ICT Deliverable D1.2, Intermediate requirements, architecture and topology report, October 2008, available at www.ict-omega.eu/publications/deliverables.html.

[OME 09] Omega ICT Deliverable D4.3, Optical wireless MAC-LLC specification – OWMAC, November 2009, available at www.ict-omega.eu/publications/deliverables.html.

[OME 10a] Omega ICT, document D4.2a, Physical layer design and specification – smart wireless optics, 2010, available at www.ict-omega.eu/publications/deliverables.html.

[OME 10b] Omega ICT Deliverable D4.5, Performance report for SWO – smart wireless optic system, June 2010, available at www.icto-mega.eu/publications/deliverables.html.

[OME 11a] Omega ICT, Projet européen, 2011, available at www.ict-omega.eu/.

[OME 11b] Omega ICT Deliverable D4.6a, Final evaluation report, March 2011, available at www.ict-omega.eu/publications/deliverables.html.

[ORA 11] ORANGE, Prediction FSO, Application d'aide à la décision pour l'implémentation de FSO pour les réseaux métropolitains à haut débit, 2011, available at www.orange.com/fr_FR/innovation/licences_logiciels/ficheslogiciels/P01258.jsp.

[PAH 95] PAHLAVAN K., LEVESQUE A.H., *Wireless Information Networks*, Wiley & Sons, New York, USA, 1995.

[PAP 95] PAPINI O., WOLFMANN J., "Algèbre discrète et codes correcteurs", *Mathématiques et Applications*, vol. 20, Springer-Verlag, 1995.

[PAR 01] PARAND F., FAULKNER G.E., O'BRIEN D.C., EDWARDS D.J., "An optical wireless test-bed system using a multiple source transmitter and a segmented receiver to achieve signal tracking", *Proceedings of SPIE, Optical Wireless Communications IV*, vol. 4530, 2001.

[PER 97] PEREZ-JIMENEZ R., BERGES J., BETANCOR M.J., "Statistical model for the impulse response on the infrared indoor diffuse channels", *Electronics Letters*, vol. 33, no. 15, pp. 1298–1301, 1997.

[PER 98a] PEREZ-JIMENEZ R., LOPEZ-HERNANDEZ F.J., SANTAMARIA A., "Monte Carlo simulation for the impulse response on diffuse IR wireless indoor channels", *Electronics Letters*, vol. 34, no. 12, pp. 1260–1261, 1998.

[PER 98b] PEREZ-JIMENEZ R., LOPEZ-HERNANDEZ F.J., SANTAMARIA A., "Modified Monte Carlo scheme for high efficiency simulation for the impulse response on diffuse IR wireless indoor channels", *Electronics Letters*, vol. 34, no. 19, pp. 1819–1820, 1998.

[PET 72] PETERSON W.W., WELDON E.J. JR., *Error-Correcting Codes*, 2nd ed., MIT Press, Cambridge, MA, USA, 1972.

[PHA 98] PHANG K., JOHNS D.A., "A 3-V CMOS optical preamplifier with DC photocurrent rejection", *Proceedings of the IEEE International Symposium on Circuits and Systems – ISCAS '98* (Cat No. 98CH36187), New York, USA, 1998.

[PHO 75] PHONG B.T., "Illumination for computer generated pictures", *Communications of the ACM*, vol. 18, 1975.

[POP 09] POPOOLA W., GHASSEMLOOY Z., AWAN M.S., LEITGEB E., "Atmospheric channel effects on terrestrial free space optical communication links", *ECAI, International Conference*, 3rd ed., Pitesti, Romania, 3–5 July 2009.

[PRO 00] PROAKIS J., *Digital Communications*, 4th ed., McGraw-Hill, New York, USA, 2000.

[PYN 94] PYNDIAH R., GLAVIEUX A., PICART A., JACQ S., "Near optimum decoding of product codes", *Global Telecommunications Conference, Globecom '94*, vol. 1, San Fransisco, USA, pp. 339–343, November 1994.

[QOF 11] Qualité Optique sans Fils, Indoor: QOFI software freeware, 2011, available at http://foton.enssat.fr/projets/Techimages/Techimages_en.php or http://departements.enst-bretagne.fr/sc/recherche/techimages/.

[REE 60] REED I.S., SOLOMON G., "Polynomial codes over certain finite fields", *Journal of the Society for Industrial and Applied Mathematics*, vol. 8, pp. 300–304, 1960.

[REI 09] REIDENBACH H.D., "Comparison of afterimage formation and temporary visual acuity disturbance after exposure with relatively low irradiance levels of laser and LED light", *ILSC*, Orlando, USA, 2009.

[SCH 96] SCHNEIER B., *Applied Cryptography*, 2nd ed., Wiley & Sons, New York, 1996.

[SCH 06] SCHMITT N.P., PISTNER T., VASSILOPOULOS C., MARINOS D., BOUCOUVALAS A.C., NIKOLITSA M., AIDINIS C., METAXAS G., "Diffuse wireless optical link for aircraft intra-cabin passenger communication", *CNSDSP*, Patras, Greece, 2006.

[SHA 48] SHANNON C.E., "A mathematical theory of communication", *Bell System Technical Journal*, vol. 27, pp. 379–423, 623–656, July and October 1948.

[SHE 79] SHETTLE E.P., FENN R.W., Models for the aerosols of the lower atmosphere and the effects of humidity variations on their optical properties, AFGL-TR-79-0214, Air Force Geophysical Laboratory, Bedford, MA, USA, 1979.

[SHE 89] SHETTLE E.P., "Models of aerosols, clouds and precipitation for atmospheric propagation studies", *Atmospheric Propagation in the UV, Visible, IR and MM Wave Region and Related Systems Aspects AGARD Conference Proceeding*, vol. 454, no. 15, pp. 1–13, 1989.

[SHI 99] SHIU D.S., KAHN J.M., "Differential pulse-position modulation for power-efficient optical communication", *IEEE Transactions on Communications*, vol. 47, no. 8, August 1999.

[SIN 01] SINGH S., COQUERET C., *Histoire des codes secrets*, Poche, Paris, 2001.

[SIV 03] SIVABALAN A., JOHN J., "Modelling and simulation of indoor optical wireless channels: a review ", *Conference on Convergent Technologies for Asia-Pacific Region (TENCON 2003)*, Bangalore, India, vol. 3, pp. 1082–1085, 2003.

[SMY 95] SMYTH P.P., EARDLEY P.L., DALTON K.T., WISELY D.R., MCKEE P., WOOD D., "Optical wireless: a prognosis", *Proceedings of SPIE Conference on Wireless Data Transmission*, Philadelphia, USA, vol. 2601, pp. 212–225, 23–25 October 1995.

[SOD 09] SODNIK Z., PERDIGUES ARMENGOL J., CZICHY R.H., MEYER R., "Adaptive optics and ESA's optical ground station", *SPIE Free Space Optic*, Paper 7464-5, 2009.

[SPA 09] SPAARMANN S., "Opportunities for a sustainable communications technology", Chancen, October 2009, available at www.hese-project.org/de/emf/WissenschaftForschung/Spaarmann_Dr.%20rer.%20nat._Stefan/4.11.9%20Chancen%20English.pdf.

[STO 09] STOTTS L.B., STADLER B., HUGHES D., KOLODZY P., PIKE A., YOUNG D.W., SLUZ J., JUAREZ J., GRAVES B., DOUGHERTY D., DOUGLASS J., MARTIN T., "Optical communications in atmospheric turbulence", *Proceedings of SPIE*, vol. 7464, 2009.

[TAN 00] TANAKA Y., HARUYAMA S., NAKAGAWA M., "Wireless optical transmissions with white colored DEL for wireless home links", *Proceedings of 11th International Symposium on Personal, Indoor and Mobile Radio Communication*, London, UK, 2000.

[TAN 02] TANAKA Y., A study on optical wireless communication systems and their applications, PhD Dissertation, Keio University, 2002.

[TEC 08] TECHIMAGES, "LD4.1: SP4: Wlan en optique en espace libre, Environnement Domestique, Typologie et Perturbateurs Optiques, Version 1.0", French regional cluster "Images et Réseaux", 2008, available at www.images-et-reseaux.com/.

[TEC 11] TECHIMAGES, French regional cluster "Images et Réseaux", 2011, available at www.images-et-reseaux.com/.

[TEM 96] Temmar A., Praseuth J.P., Palmier J.F., Scavennec A., "Photodiode de type MSM sur substrat d'InP ", *Journal de physique III*, 1996.

[TOU 11] TOUTAIN L., Local Networks and the Internet, ISTE, London, John Wiley & Sons, New York, 2011.

[TRÉ 67] TRÉHEUX M., "Transmission par laser", *Radome*, vol. 10, pp. 17, 1967, available at www.apast.asso.fr/docs_du_site/Radome_10.pdf.

[VAS 80] VASSALO C., *Electromagnétisme classique dans la matière*, Dunod, Paris, 1980.

[VAS 97] VASSEUR H., OESTGES C., VANDER VORST A., "Influence de la troposphère sur les liaisons sans fil aux ondes millimétriques et optiques", *Propagation électromagnétique du décamétrique à l'angström*, 3es journées, Rennes, France, 1997.

[WAI 05] WAINRIGHT E., REFAI H.H., SLUSS J.J., "Wavelength diversity in free space optics to alleviate fog effects", *Proceedings of SPIE*, vol. 5712, 2005.

[WAL 09] WALTHER F.G., MOORES J.D., MURPHY R.J., MICHAEL S., NOWAK G.A., "A process for free-space laser communications system design", *Proceedings of SPIE*, vol. 7464, 2009.

[WEI 08] WEICHEL H., "Laser beam propagation in the atmosphere", POTTER R.F. (ed.), *The International Society for Optical Engineering*, Bellingham, Washington, vol. TT-3, pp. 60–63, 1989.

[WEI 71] WEINSTEIN S.B., EBERT P.M., "Data transmission by frequency-division multiplexing using the discrete Fourier transform", *IEEE Transactions on Communications*, vol. 19, pp. 628–634, October 1971.

[WIS 97] WISELY D., NEILD I., "A 100 Mbit/s tracked optical telepoint", *Proceedings of the Eighth IEEE International Symposium on Personal Indoor and Mobile Radio Communications* (PIMRC'97), Helsinki, Finland, vol. 1–3, pp. 964–968, 1997.

[WOL 03] WOLF M., KRESS D., "Short-range wireless infrared transmission: the link budget compared to RF", *IEEE Wireless Communications Magazine*, pp. 8–14, April 2003.

[WOL 08] WOLF M., The new standard 60825-1:2007, edition 2 for safety of laser products – safety requirements for class 1 sources, internal OMEGA WP4 document, April 2008.

[WOL 09] WOLF M., GROBE L., LI J., Demo V2 link budget and things to be done, internal presentation, European Project Omega, November 2009.

[WON 07] WON E.T., "Visible light Communication SG", *IEEE 802.15.3*, Atlanta, USA, 2007.

[YAN 00] YANG H., LU C., "Infrared wireless LAN using multiple optical sources", *IEEE Proceedings Optoelectronics*, vol. 147, 2000.

[YUN 92] YUN G., KAVEHRAD M., "Spot-diffusing and fly-eye receivers for indoor infrared wireless communications", *IEEE International Conference on Selected Topics in Wireless Communications, Vancouver Conference Proceedings*, Vancouver, Canada, pp. 262–265, 1992.

[ZIE 98] ZIEMER R.E., TRANTER W.H., FANNIN D.R., *Signals and Systems: Continuous and Discrete*, 4th ed., Prentice Hall, Englewood Cliffs, 1998.

[ZYS 10] ZYSS J., "Lasers polymers and green lasers", *Workshop Lasers et communication*, Institut Télécom, Paris, December 2010.

List of Figures

Figure 1.1. Stele of the Lady Taperet (Louvre museum) 2
Figure 1.2. Device and Newton's rings 3
Figure 1.3. Emission and absorption of a photon 5
Figure 1.4. The International Congress of Physics,
Solvay in Brussels (1927) 6
Figure 2.1. Aeneas the Tactician device 9
Figure 2.2. Roman telegraph station (low-relief of the
Trojan column in Roma) 10
Figure 2.3. Robert Hooke's telegraph 11
Figure 2.4. The Chappe's system used high points
(towers, bell towers, etc.) 12
Figure 2.5. Mechanism developed by Chappe: three articulated arms 13
Figure 2.6. The Chappe's telegraph [JOI 96] 15
Figure 2.7. Signals used in the final code of Chappe's telegraph 16
Figure 2.8. The Chappe's code [Libois, unpublished] 17
Figure 2.9. An English heliograph 19
Figure 2.10. Lesuerre's heliograph 19
Figure 2.11. Mangin's telegraph 20
Figure 2.12. Using the optical telegraph of Colonel Mangin 21
Figure 2.13. The photophone 21
Figure 2.14. Marconi radio equipment 22
Figure 3.1. Communication history 26
Figure 3.2. Modules of an optical wireless device 26
Figure 3.3. Line of sight 27
Figure 3.4. (a) Multisectoral, (b) angular diversity, and
(c) imaging reception 28
Figure 3.5. Wide line of sight 29
Figure 3.6. Diffusion 30
Figure 3.7. Controlled diffusion 31

Figure 3.8. Representation of the real components of the electric
and magnetic fields along the direction of propagation   35
Figure 3.9. Free-space propagation with reflection   39
Figure 3.10. Local approximation of a wave surface by a plane wave   42
Figure 3.11. Electromagnetic spectrum   43
Figure 3.12. Spectral band   44
Figure 3.13. Curve of brightness of solar radiation   48
Figure 3.14. The optical spectrum of a tungsten filament lamp   49
Figure 3.15. Spectrum of quasi-monochromatic source (v_0)   49
Figure 3.16. OSI model example   51
Figure 3.17. TCP/IP model   52
Figure 3.18. Communication between SPOT-4 and Artemis with
Silex laser system (Source: ESA)   54
Figure 3.19. Visplan (Source: JVC)   58
Figure 3.20. Wireless optic ecosystem   61
Figure 4.1. General form of narrowband signal   65
Figure 4.2. Photo electrical detection and wavelength   66
Figure 4.3. Equivalent baseband model   67
Figure 4.4. Example of optical disruptors   71
Figure 4.5. Multipath   73
Figure 4.6. Example of overlapping symbols   74
Figure 4.7. Example of impulse response in confined environment   74
Figure 4.8. Example of impulse response $h(t)$ for two FOV values   75
Figure 4.9. Experimental reflection patterns of a rough cement surface
(before and after white painting)   77
Figure 4.10. Reflection patterns for a varnished wood surface   80
Figure 4.11. Schematic view of the modified Monte Carlo model   82
Figure 4.12. Subdivision of a 3D cube into patches   83
Figure 5.1. Transmittance of the atmosphere due to molecular absorption   89
Figure 5.2. Specific attenuation (dB/km) due to the rain   97
Figure 5.3. Wet snow: attenuation in function of snowfall   98
Figure 5.4. Dry snow: attenuation in function of snowfall   98
Figure 5.5. Influence of a large turbulent cell (deviation)   99
Figure 5.6. Influence of a small turbulent cell (widening of the beam)   99
Figure 5.7. Heterogeneities (scintillations)   100
Figure 5.8. Variation of the attenuation due to scintillation   101
Figure 5.9. Synoptic view of the experimental set-up   103
Figure 5.10. Light beam of the transmissometer and laser   104
Figure 5.11. Measurements and comparison with Kruse's model   105
Figure 5.12. Measurements and comparison with Kim's model   106
Figure 5.13. Measurements and comparison against Al Naboulsi's model
(advection)   106

Figure 5.14. Measurements and comparison against Al Naboulsi's model
(convection) 107
Figure 5.15. Distribution in France of the number of days
with fog per year. 109
Figure 5.16. RVR variations observed on the site of La Turbie 110
Figure 5.17. Direct beam transmissometer 111
Figure 5.18. Reflected beam transmissometer 111
Figure 5.19. Emission part of a transmissometer installed on the
site of La Turbie. 112
Figure 5.20. Schematic representation of the visibility measurement
by backscatter 112
Figure 5.21. Schematic representation of the measurement of forward
scatter visibility 113
Figure 5.22. Example of a scatterometer implemented on a motorway area. . . . 113
Figure 5.23. Input data acquisition screen. 115
Figure 5.24. Results presentation screen. 115
Figure 5.25. Screen showing the wave front propagation. 116
Figure 6.1. Distance as a function of the divergence 121
Figure 6.2. Schematization of the elements of calculation of the
optical power. 122
Figure 6.3. Emission diagram and half-power angle 124
Figure 6.4. Example of emission diagram. 125
Figure 6.5. Example of emission (T_x) and reception (R_x) devices 126
Figure 6.6. Principle of retroreflexion 133
Figure 6.7. Diagram of radiation of the emitted and retroreflected beam 134
Figure 6.8. Emitter–receptor link. 136
Figure 6.9. Profile of the impulse response for case 4 (LOS + DIF) 138
Figure 7.1. Schematic cross section of an eye 142
Figure 7.2. Low size and wide sources 146
Figure 7.3. Configuration of measurement 147
Figure 7.4. Allowed radiation intensity class 1 [WOL 08] 148
Figure 8.1. Diagram of an optical wireless transmitter 158
Figure 8.2. Diagram of an optical wireless receiver. 158
Figure 8.3. Band diagram of a crystal of semiconductor material. 159
Figure 8.4. LED spontaneous emission structures 161
Figure 8.5. Lambert's law: (a) $I(\theta) = I(0).cos\theta$ and (b) $I(\theta) = I(0).(cos\theta)n$. 162
Figure 8.6. Multi-chip LED spectra. 163
Figure 8.7. Phosphor-coated LED spectra. 163
Figure 8.8. Laser structure: thin epitaxial layers, stripe
and Fabry–Perot resonator. 164
Figure 8.9. Fabry–Perot and DFB lasers. 165
Figure 8.10. Photovoltaic cell. 167
Figure 8.11. Responsivity versus wavelength 168

Figure 8.12. Optical emission 171
Figure 8.13. Cassegrain telescope 171
Figure 8.14. Example of optical transmitter. 171
Figure 8.15. Example of aspherical lens (a) and Fresnel (b) 173
Figure 8.16. Fisheye lens [ALH 06] 174
Figure 8.17. Example of optical high-pass filtering 175
Figure 9.1. The different modulation techniques. 179
Figure 9.2. The different input–output techniques 183
Figure 9.3. BER of an NRZ-OOK modulation associated
with a BCH code ($t = 1$) 189
Figure 9.4. BER of an NRZ-OOK modulation associated with an RS code 190
Figure 9.5. Schematization of a convolutional encoder
with performance $Rc = k/n$ 191
Figure 9.6. Example of a schematic diagram of the convolutional
encoder CC (7.5) 192
Figure 9.7. Schematic example of error probability of
an NRZ-OOK modulation 193
Figure 10.1. 802.11 IR frame 208
Figure 10.2. OWMAC layer. 210
Figure 10.3. OWMAC frame 211
Figure 11.1. FSO between X and Y sites. 214
Figure 11.2. FSO equipment example [BOU 04] 215
Figure 11.3. FSO implementation: potential problems to avoid 218
Figure 11.4. Data site and equipment example for Paris 222
Figure 11.5. Example of results for Paris 222
Figure 11.6. Example of equipment profile for Paris 223
Figure 11.7. Example of architecture of a wireless radio system
like WiFi 227
Figure 11.8. Example of architecture of a wireless optical system 228
Figure 11.9. Simulation in a reference room 230
Figure 11.10. Example of a furnished room 231
Figure 11.11. "A" configuration example 232
Figure 11.12. "B" configuration example 232
Figure 11.13. "C" configuration example 233
Figure A1.1. Snell–Descartes' laws 245
Figure A1.2. Real object – real image 246
Figure A1.3. Real object – virtual image 246
Figure A1.4. Solid angle 247
Figure A1.5. Pulsed mode 251
Figure A1.6. Luminance of the source. 253
Figure A1.7. Geometrical extent of a light beam. 254

List of Tables

Table 3.1. Prefixes commonly used in electromagnetism. 44
Table 3.2. Frequencies, periods, and corresponding wavelengths
of different electromagnetic spectra 47
Table 3.3. Historic list of wireless optical communications 56
Table 3.4. Comparison between 60 GHz radio and wireless
optical technologies. 60
Table 4.1. Reflection coefficient ρ 78
Table 4.2. Reflection coefficient of different materials 80
Table 5.1. Parameters characterizing the particles size distribution. 92
Table 5.2. Parameters "a" and "b" for wet and dry snow. 97
Table 5.3. International visibility code. 102
Table 5.4. Annual frequency of fog events in Belgium 108
Table 6.1. Half-power angles and their respective m parameters 124
Table 6.2. Optical budget, typical values 135
Table 6.3. Five typical optical budget examples 137
Table 6.4. Typical values of power of ambient noise and SNR 140
Table 7.1. Classification of the lasers 144
Table 7.2. Legal mention of information 145
Table 7.3. Energy comparison. 153
Table 8.1. Comparative LED–LD 166
Table 9.1. Mean power and bandwidth 181
Table 10.1. RC5 code for DVD reader. 202
Table 10.2. Comparison protocol OSI versus IrDA 203
Table 10.3. IrDA link performance parameters. 205
Table 10.4. Wireless optical link protocols 212
Table 11.1. Some FSO manufacturers 217
Table 11.2. Values of geometrical attenuation for various distances 219
Table 11.3. Link margin of three typical systems for a distance of 500 m. 220
Table 11.4. Availability results of selected links. 224

Table 11.5. Distribution of room surfaces. 226
Table 11.6. Cumulative distribution of surfaces 229
Table 11.7. Maximum angle and distance. 230
Table 11.8. Objective of system characteristics 231
Table 11.9. Calculated geometric attenuation 233
Table A1.1. Examples of refractive indexes 243
Table A1.2. Usual illumination values. 248
Table A1.3. Usual luminance values. 249
Table A1.4. Equivalence table between radiometry and photometry 255

List of Equations

Equation 3.1. Maxwell's equations (unspecified medium) 32
Equation 3.2. Conservation of charge relation 32
Equation 3.3. Equations for a free-space propagation 33
Equation 3.4. Electric field (1st equation) 33
Equation 3.5. Electric field (2nd equation) 33
Equation 3.6. Maxwell equations (isotropic and linear
homogeneous medium) 34
Equation 3.7. Propagation equations 34
Equation 3.8. Electric and magnetic field 34
Equation 3.9. Phase of the wave 35
Equation 3.10. Phase velocity 36
Equation 3.11. Phase velocity in the vacuum 36
Equation 3.12. Electromagnetic energy in a volume V 36
Equation 3.13. Energy density 36
Equation 3.14. Electromagnetic energy density 37
Equation 3.15. Energy flux 37
Equation 3.16. Poynting vector 37
Equation 3.17. Average energy value 37
Equation 3.18. Total field from two sources 39
Equation 3.19. Average intensity of the total field 39
Equation 3.20. Visibility from two optical paths 40
Equation 3.21. Equations of an optical ray 40
Equation 3.22. Vector normal to the wave front 41
Equation 3.23. Planck's law of the Sun 48
Equation 4.1. Narrowband signal (1st representation) 64
Equation 4.2. Narrowband signal (2nd representation) 64
Equation 4.3. Received useful signal 65
Equation 4.4. Equivalent baseband model 67
Equation 4.5. Transmitted average optical power 68

Equation 4.6. Received average optical power 68
Equation 4.7. Direct current channel gain 68
Equation 4.8. Average electrical power of the useful signal 69
Equation 4.9. Noise power expression 69
Equation 4.10. Spectral density of dominant noise 70
Equation 4.11. Electric SNR 71
Equation 4.12. Electrical SNR in wireless optical communication 72
Equation 4.13. Impulse response model 75
Equation 4.14. Rugosity criteria 76
Equation 4.15. Lambert's relation 78
Equation 4.16. Example of Lambert's reflection 78
Equation 4.17. Phong's relation 79
Equation 5.1. Beer's law 88
Equation 5.2. Extinction coefficient 88
Equation 5.3. Molecular scattering coefficient 90
Equation 5.4. Molecular absorption coefficient 90
Equation 5.5. Aerosolar scattering coefficient 91
Equation 5.6. Distribution of particle size 92
Equation 5.7. Kruse and Kim model 94
Equation 5.8. Attenuation by an advection fog 95
Equation 5.9. Attenuation by a convection fog 96
Equation 5.10. Rain attenuation 96
Equation 5.11. Attenuation due to snow 97
Equation 5.12. Scintillation variance 100
Equation 6.1. Elementary annular angle 122
Equation 6.2. Light intensity 122
Equation 6.3. Radiated total power 123
Equation 6.4. Intensity radiated in the normal direction 123
Equation 6.5. Determination of the angle HP 123
Equation 6.6. Determination of the value m 123
Equation 6.7. $P(\varphi)$ power 124
Equation 6.8. Gain in the case of a non-imaging concentrator 126
Equation 6.9. Effective surface (1st representation) 126
Equation 6.10. Effective surface (2nd representation) 127
Equation 6.11. Rectangular function 127
Equation 6.12. Limit angle of reception 127
Equation 6.13. Theoretical sensitivity of the photodiode 128
Equation 6.14. Relation between P_t and P_r 128
Equation 6.15. Relation between P_r and I 129
Equation 6.16. Received power 129
Equation 6.17. Linear geometrical loss in line of sight 129
Equation 6.18. Response of the impulse of the channel 130
Equation 6.19. Margin of the system 130

Equation 6.20. Coverage surface 130
Equation 6.21. Gain DC of the channel 132
Equation 6.22. Received power. 134
Equation 6.23. Simplified received power 135
Equation 6.24. Binary error rate 139
Equation 8.1. Cutoff wavelength 159
Equation 8.2. Emission wavelength 159
Equation 8.3. Current of the photodiode. 168
Equation 8.4. Multiple primary photocurrent proportional
to the incident radiant power 169
Equation 9.1. Number of M states symbol 178
Equation 9.2. Coding efficiency 186
Equation 9.3. BCH codes 188
Equation 9.4. RS codes. 189
Equation 11.1. Geometric attenuation 219
Equation 11.2. FSO margin 220
Equation A1.1. Refractive index 243
Equation A1.2. Solid angle 247
Equation A1.3. Monochromatic luminous flux. 249
Equation A1.4. Polychromatic luminous flux 250
Equation A1.5. Global energy flux. 250
Equation A1.6. Bouguer's relation 250
Equation A1.7. Average power in pulsed mode 251
Equation A1.8. Power (dBm) 252
Equation A1.9. Intensity of a source 252
Equation A1.10. Energy luminance 253
Equation A1.11. Illumination of receiving surface E 253
Equation A1.12. Light beam. 254

Index

A

absorption, 5
aerosols, 87
availability, 268

C, D

cryptography, 149-151
diffusion, 30, 84, 113
DIV, 231

E

emitter, 171, 130
extinction, 18, 88, 91, 94, 95, 110
eye, 142

F

fading, 60, 65, 66, 183
fog, 107

G, H, I

GaAs, 160, 169
Graham Bell, 7, 20
heliograph, 18
impulse response, 75
InP, 271

J, K, L

laser, 54, 104, 262, 264, 267,
   269, 272
legislation, 141, 154
link margin, 179, 216, 218,
   221, 223
LOS, 27, 56, 230, 262, 265

M, N

Mangin, 20, 21
Maxwell equation, 32
model, 50, 51, 52, 105
modulation, 178, 179, 180,
   189, 190, 193, 261, 265,
   267
noise, 71

P, Q

photoreceptor, 66, 70
propagation model, 63, 64, 81, 116
Quality of Service, 139

R

receiver, 130, 132, 136
reflection, 39, 76, 78, 132
refraction, 36

S

safety, 141, 149, 216
scintillation, 100, 220
spectral, 4, 39, 44-47, 49, 57, 58,
    67, 69-72, 91, 128, 139, 148,
    155, 160, 161, 164, 166, 176,
    178, 179, 182, 199, 206, 238,
    239, 249, 250
standard, 50, 51, 58-60, 142-144,
    146, 147, 149, 153, 177, 180, 183,
    187, 192, 199, 205, 207-209, 211,
    226, 238, 249

U, W

unit, 43-47, 50, 87, 88, 90-92,
    121, 124, 127, 148, 170, 179,
    197, 207, 211, 247-249, 251-253,
    255, 257
wavelength, 94, 159, 168, 174, 199
WLOS, 27, 29, 56, 262